T0130715

Are Ships Different?

Policies and Procedures for the
Acquisition of Ship Programs

Jeffrey A. Drezner, Mark V. Arena, Megan McKernan,
Robert Murphy, Jessie Riposo

Prepared for the Office of the Secretary of Defense and the United States Navy

NATIONAL DEFENSE RESEARCH INSTITUTE

The research described in this report was prepared for Office of the Secretary of Defense (OSD) and the United States Navy. The research was conducted in the RAND National Defense Research Institute, a federally funded research and development center sponsored by OSD, the Joint Staff, the Unified Combatant Commands, the Navy, the Marine Corps, the defense agencies, and the defense Intelligence Community under Contract W74V8H-06-C-0002.

Library of Congress Control Number: 2011940470

ISBN: 978-0-8330-5013-7

The RAND Corporation is a nonprofit institution that helps improve policy and decisionmaking through research and analysis. RAND's publications do not necessarily reflect the opinions of its research clients and sponsors.

RAND® is a registered trademark.

Published 2011 by the RAND Corporation
1776 Main Street, P.O. Box 2138, Santa Monica, CA 90407-2138
1200 South Hayes Street, Arlington, VA 22202-5050
4570 Fifth Avenue, Suite 600, Pittsburgh, PA 15213-2665
RAND URL: http://www.rand.org/
To order RAND documents or to obtain additional information, contact
Distribution Services: Telephone: (310) 451-7002;
Fax: (310) 451-6915; Email: order@rand.org

Preface

This research seeks to improve the policies and procedures used by the U.S. Department of Defense (DoD) and the U.S. Navy in providing and supporting oversight of ship acquisition programs. The focus of the research was to (a) identify any aspects of major ship acquisition programs that appear to deviate substantially from the 5000-series management process, (b) identify ambiguities in that process for ship programs, and (c) suggest changes in either DoD or Navy policies and procedures that could ameliorate any undesirable consequences (such as reporting requirements that are not useful to Program Managers or to oversight) and resolve procedural uncertainty for major ship acquisitions.

This monograph presents the results of research that was conducted between March 2008 and February 2009. It is intended for an audience who has some background in defense acquisition policy. Particularly, it assumes that the reader is familiar with and has a basic understanding of the DoD 5000-series acquisition regulations and instructions.

During the course of this research, the main DoD acquisition regulation, DoD Instruction (DoDI) 5000.2 (May 2003), and Secretary of the Navy Instruction (SECNAVINST) 5000.2C were updated and reissued as DoDI 5000.02 (December 2008) and SECNAVINST 5000.2D (October 2008). Our direct analysis focuses on the most recent versions of the regulations, but our interviews of stakeholders were done while the older instructions were in force. It was not possible to reinterview all the stakeholders. Nevertheless, the changes regarding shipbuilding in either document were not substantial, were already known, or existed as other notes or instructions. We feel, therefore, that our interview findings remain relevant.

This research was sponsored jointly by Program Executive Office (PEO) Ships and the Office of the Under Secretary of Defense (Acquisition, Technology, and Logistics) Portfolio Systems Acquisition—Naval Warfare and conducted within the Acquisition and Technology Policy Center of the RAND National Defense Research Institute, a federally funded research and development center sponsored by the Office of the Secretary of Defense, the Joint Staff, the Unified Combatant Commands, the Navy, the Marine Corps, the defense agencies, and the defense Intelligence Community.

For more information on the Acquisition and Technology Policy Center, see http://www.rand.org/nsrd/ndri/centers/atp.html or contact the director (contact information is provided on the web page).

Contents

Figures

Tables

Summary

The management and oversight of a major defense acquisition program are exceedingly complex processes that must balance and reconcile diverse interests and differing perspectives and constituencies. Program Managers might focus more on near-term goals, including that of getting the new capability into the hands of operational users as quickly as possible. Others might place greater emphasis on minimizing risk by insisting on extensive testing before production starts. Still others are responsible for ensuring that all funds are expended in ways consistent with law and congressional intent and focus their efforts accordingly.

The U.S. Department of Defense has a well-established set of policies, procedures, and organizations for acquisition program management and oversight, described in the 5000 series of directives and instructions. These documents describe procedures and organizational responsibilities for program management, major milestones and key technical reviews, systems engineering, and test and evaluation.

Not all weapon systems fit comfortably within this framework. Indeed, every program is unique in one or more important ways. Some systems, including ships, have no dedicated full-scale test units; rather, every unit produced is expected to enter service. Ships also have several other characteristics that make them unique:

- length of time to design and build
- importance of industrial/political factors
- concurrency of design and build
- complexity
- low quantity/production rate
- high unit cost
- type of funding
- test and evaluation procedures.

The formal acquisition process is intended to be flexible enough to accommodate program differences through tailoring,[1] in which program management, program

---

[1] Tailoring is described in DoDI 5000.02 as how the MDA establishes "regulatory information requirements and acquisition process procedures . . . to achieve cost, schedule, and performance goals."

executive office, and Office of the Secretary of Defense (OSD) oversight officials craft a management and oversight approach that accommodates unique program characteristics but still satisfies statutory and regulatory constraints. Nevertheless, such flexibility requires considerable personal initiative to execute. Ship acquisition personnel for both the Navy and the OSD have become increasingly frustrated that the same acquisition strategy and program issues are addressed repeatedly, both within and across programs. At the request of the Navy and the OSD, RAND researchers therefore examined current policies, interviewed current personnel on current processes, documented the extent to which process tailoring is needed for shipbuilding programs but may not be accommodated, and developed suggestions for improvement.

Current Processes and Accommodations

The current generic acquisition process revolves around three milestones and associated life-cycle phases.

Milestone A is the decision point associated with entry into the technology development phase. It is typically reached once the Analysis of Alternatives is complete and a specific technical solution is proposed. At Milestone A, the MDA[2] approves the preferred material solution, approves the technology development strategy, and prepares a certification memo as required by statute.

Milestone B typically marks the formal initiation of a program and entry into the engineering and manufacturing development (EMD) phase. By this milestone, the program usually has had a preliminary design review, demonstrated relevant technologies and manufacturing processes, and determined its cost and schedule baseline. Here, too, the MDA must prepare a certification memo as required by statute.

Milestone C typically denotes entry into the production phase and authorizes a program to begin production at a low rate. By this point in the program, engineering and manufacturing development is complete and required testing and operational assessments have been successful.

Shipbuilding programs can differ from this generic process in several ways. Following Milestone A, acquisition programs typically have a technology development phase, with system design and development waiting until formal program initiation, which occurs at Milestone B. However, shipbuilding programs can be formally initiated at Milestone A (at the discretion of the MDA), thus beginning their formal pro-

---

[2]   The MDA is the Department of Defense official responsible for making programmatic decisions during the acquisition process. The Under Secretary of Defense, Acquisition, Technology, and Logistics (AT&L) is generally the MDA for larger acquisition programs, acquisition category (ACAT) ID, and ACAT IAM. The head of a DoD component or Component Acquisition Executive is the MDA for the ACAT IC, ACAT II, and ACAT III programs.

gram activities earlier than other weapon systems. Ship programs tend to have somewhat more concurrency of technology development and system design activities.

Similarly, for shipbuilding programs, Milestone B rather than Milestone C essentially marks the start of initial production by authorizing lead ship construction. Unlike other programs, ship programs can begin manufacture during the engineering and manufacturing development phase that follows Milestone B in the form of the lead ship. For other programs, Milestone C authorizes low-rate initial production, something that has less meaning for ships because lead ship construction has already begun. For nonship programs, Milestone C is also intended to denote the completion of development and initial operational testing. Since the lead ship may not have been delivered and tested, Milestone C will necessarily have a different meaning for ship programs.

Despite variations in practice, DoDI 5000.02 is ambiguous or lacks specific language regarding how to tailor ship programs. For example, whereas Milestone B may authorize production of the lead ship, there is no corresponding language that defines when low-rate initial production occurs for ships if production begins at Milestone B. There is also no specific language for ships on full-rate production and Milestone C. The differing (and sometimes ambiguous) meaning of milestones for shipbuilding programs leads to confusion among various acquisition stakeholders. Furthermore, the Secretary of the Navy acquisition instruction is not always consistent with the DoD instruction. For example, the Navy instruction notes that Milestone B authorizes the lead ship and initial follow ships. The DoD instruction states that Milestone B typically authorizes the lead ship and that long lead[3] for follow ships may also be approved.

Stakeholder Views

To explore how the ambiguities of current guidance may affect shipbuilding programs and acquisition issues more generally, we conducted more than two dozen interviews with representatives from the Navy and from OSD.[4] These interviews covered questions on how ships differ from other major defense acquisition programs, what issues or problems arise from these differences in following the 5000 series of instructions, and what regulatory changes could facilitate the acquisition of ships.

Both OSD and Navy interviewees noted that the length of time to design and build a ship was the primary difference between ships and other acquisition programs. They also noted that politics and industrial base considerations played more prominent

---

[3] Long lead typically refers to the materials or services that must be procured in advance of construction, such as steel or propulsion equipment that is procured before the start of construction.

[4] We did not interview industry stakeholders in this study, although we have become a familiar with industry views from other research. The sponsors of this study wanted us to focus on government stakeholders. We believe that the breadth of interviews we conducted adequately covers the issues addressed in this study.

roles in shipbuilding than in other acquisition programs. Other areas in which interviewees said that shipbuilding programs are unique are the concurrency of design and build, greater complexity, high unit cost, and low production rate.

Some interviewees suggested that process tailoring[5] is sufficient to address the unique requirements of ships, with one even suggesting that all acquisition programs require tailoring. Others said that ambiguities in language made implementation of the 5000 process more difficult for ships. Among specific areas of difficulty that interviewees mentioned regarding ships were interpreting DoD instructions for ships (including initial and full-rate production), the content and timing of documentation requirements, testing and evaluation issues, statutory issues, and other policy and process issues.

Most interviewees did not think the 5000 process was irreparable, but many suggested ways to improve it to accommodate shipbuilding programs. Interviewees thought the 5000 process flexible, but many said that they found tailoring difficult. One claimed that tailoring resulted in more reviews and meetings, providing little incentive to seek it. Some said that improved guidance and less formalization of tailoring might make it more useful. Some felt that low-rate initial production and full-rate production distinctions should not apply to ships. OSD interviewees suggested more and earlier component or subsystem-level testing for ships, whereas Navy personnel suggested different language on technological development, recognizing that ships are a system of systems, and simplification of the system engineering process. Both OSD and Navy personnel felt that improving the ability to tailor, rethinking the meaning of currently ambiguous definitions (such as that for low-rate initial production and full-rate production) for ships, and rethinking the best way to test and evaluate ships would be helpful. One interviewee suggested that capturing these definitions in an annex to the 5000 process would be helpful.

Program Comparisons

To better evaluate the perceptions of stakeholders about the unique issues of shipbuilding programs, we gathered data on acquisition time lines and major program activities for several ship and nonship acquisitions.

The data confirmed that shipbuilding programs often have compressed early phases, contract awards that define program phases as well as the sequence of activities, relatively small total quantities, low annual production rates, a significant portion of the total quantity on contract before testing of the lead ship is complete, and a more significant role for the industrial base in influencing program structure and

---

[5]   *Tailoring*, as used by interviewees, refers to the ability to alter various procedures and requirements of the acquisition process to accommodate the unique aspects of a program. This is consistent with how *tailoring* is defined in regulation.

contracting activity. Perhaps most important, the data confirm an apparent mismatch between major milestones and key program events, such as contract awards and testing. Ship program design and build events do not appear to align well with the intent, timing, scope, and content of some milestone reviews. Often, contract awards denote different design stages (e.g., system design, functional design, contract design) before Milestone B (or II) rather than technology development and demonstration. Milestone B tends to not only approve continued design activity (detail design) but also initial production (e.g., the lead ship). Under DoDI 5000.02, Milestone B is intended as the start of product development—integrating technologies and maturing concepts into a form intended for deployment to the warfighter—whereas ship programs tend to treat this milestone as a continuation of design activities. With some notable exceptions, system and subsystem demonstration through testing must wait for delivery and test of the lead ship, as opposed to the construction of developmental systems intended only for test. Ship programs also differed from each other, with the exact sequence, timing, and scope of contract awards varying by ship type, size, maturity of design and technology, the roles and responsibilities of government and industry, the preferred design tools, and the characteristics of the industrial base for each program.

We also identified differences between ship and other programs that appear to be real but do not have clear metrics. Milestone decisions and technical activities may require information at a level of detail not available until later stages of a ship program. Because the lead ship is intended as a deployable asset, live fire test activities may, because of the risk of damage, be inappropriate for ships, although this problem is becoming increasingly prevalent in programs with high unit costs. Although some major acquisition programs, such as satellites, may seem similar to ships in their long design and build time frame, deployed and operational first units, small quantities, and low production rates, they can also differ in ways, such as the workforce size and skill set needed to develop them and their operational environments. This underscores the need to tailor acquisition processes.

Yet care must be taken in tailoring programs. For example, because annual production rates for many complex ships are low and steady, the normal distinction between low-rate and full-rate production made by Milestone C is not relevant. Simply dropping Milestone C for ships, however, risks losing other attributes for it that are relevant to oversight, such as the completion of development and initial testing.

Designing the Ideal Ship Acquisition Process and Strategy

How might acquisition processes and milestones be best tailored for ship programs? To gain insight, RAND researchers considered two hypothetical possibilities: an unconstrained process in which constraints and requirements can be ignored to understand the range of possible alternatives, and a constrained case in which constraints affect the timing and scope of oversight activities.

In the unconstrained case, Milestone A, as the start of major activity, is relatively fixed, but some latitude is available for technical and engineering activities and subsequent milestones and other oversight activities. Milestone B may be as early as the beginning of detail design work or as late as the start of construction for the lead ship. Milestone C could occur as early as the start of construction for the first ship or as late as the completion of initial operational test and evaluation.

This unconstrained case suggests a number of desirable "best" practices. The shipyards, weapon-system contractors, and Naval Sea Systems Command (NAVSEA) should collaborate, beginning with feasibility studies. Lead responsibility would shift, depending on activity, life-cycle phase, and relative competency. Prototyping (with or without competition) would be done to the maximum extent possible at the component and subsystem level. Early and continuous developmental and operational testing would be performed. Verification through inspection, analysis, modeling and simulation, similarity, and demonstration would all be acceptable practices. The gap between lead ship and follow ship would increase to reduce technical and operational risk; the follow ship would be built only after initial operational test and evaluation (IOT&E) is complete.

There is no "right" alternative; each has pros and cons. In a case with late milestones and little overlap of program phases, the technology and program baseline are more mature when decisions are made. Rework and redesign risks are minimized. However, longer program durations may lead to requirements "creep" and pose a significant challenge to the industrial base. Waiting to procure the follow ship until after the lead ship has completed IOT&E may be impractical, resulting in an excessive production gap, learning loss, higher material costs, and vendor base impacts. Where Milestone B can be aligned as needed and there is some concurrency in design and construction, the milestones can be aligned with the key functional activities of the ship design/build process. This option will require explicit process tailoring to define when Milestone B occurs for each ship program. An early milestone case with overlap of technology development, design, and construction phases would potentially allow mature technologies to be fielded more rapidly. These milestones mark the start of key functional activities in the ship design/build process. Nevertheless, satisfying the documentation requirements earlier in the process may be very challenging or might require a waiver. The concurrency of the process introduces risk of various kinds. The "optimal" program structure is thus very closely tied to the characteristics of a particular ship concept (e.g., technological maturity, design maturity, relevant industry capabilities) and an acceptable balancing of associated risks.

In the constrained case, however, the programmatic choices the Navy can make are influenced by a large set of factors, including technical and engineering activities, statutory and regulatory requirements, industrial base issues (workforce, financial viability of shipyards), capital equipment requirements, force structure requirements, political factors, and overall fiscal constraints. As a result, the constrained oversight case looks like the nominal process for current ship programs. Milestone B denotes the start of detailed design and authorizes lead-ship construction, with an interim progress review authorizing initial follow ships. The role of Milestone C remains unclear, especially when few ships are to be built in a program. Milestone C might replace the interim progress review that authorizes follow ship construction. However, stakeholders had mixed opinions on this; some supported the idea, but others said that moving Milestone C would provide only limited oversight value because much information would remain unchanged after Milestone B.

Policy Options

There is a range of policy options for ship programs to reconcile the problems posed by unique characteristics of shipbuilding programs and ambiguities in DoD instructions. At one extreme, policymakers may choose to exempt ship programs from the DoD 5000 series. At another extreme, the DoD instructions might be rewritten to include language for each commodity type.

Exempt Ship Programs from the DoD Instructions

Exempting ship programs from the DoD instructions would give the Navy increased flexibility to design and manage ship programs. This would effectively shift program oversight to the Navy. Yet it would also shift many of the same problems to the Navy, given the ways shipbuilding programs can differ from each other.

Remove All References to Commodity Types in the DoD Instructions

A less extreme measure would remove all references to commodity types in the DoD instructions. Indeed, ships and satellite programs are the only weapon systems currently mentioned explicitly. Removing such explicit mention would leave just high-level guidance to tailor processes as appropriate. However, this option would not address the core issues that pose real challenges to ship programs. In particular, stakeholders would still debate what process tailoring is required for each ship program because of which characteristics, with no additional guidance on the range of acceptable options.

Clarify the Language and Interpretation for Ship Programs

Clarifying the language and interpretation for ship programs could help resolve ambiguities and conflicts in requirements. Additional guidance on how the process could be adjusted (tailored) for ship programs could ensure a more standardized interpretation of regulations and help coordinate the efforts of Program Managers and oversight officials. A policy memo could make explicit those parts of the acquisition process that need to be tailored for ship programs as well as the range of tailoring options available. Each ship program would need to address each tailoring area as part of its acquisition strategy documentation.

Rewrite the Base Acquisition Regulation to Include Language for Each Weapon Type

Going beyond some clarifying language to rewrite the base acquisition regulation to include language for each weapon type could result in new problems rather than solving existing ones. Rewriting the base acquisition regulation could reduce program management flexibility and at the same time result in differing processes for ships, satellites, launch vehicles, armored vehicles, aircraft, and other programs, ultimately resulting in a number of completely different, independent acquisition processes. The result would be a highly complex set of acquisition regulations and processes adding to the burden of both Program Managers and oversight officials.

Conclusions and Recommendation

In the near term, we recommend clarifying the language and interpretation of existing regulations and guidance. This would involve making the language in DoD instructions more internally consistent and broader to mitigate the most critical ambiguities, aligning the language and intent of DoD instructions with those of the Secretary of the Navy, and providing more specific guidance on a standardized interpretation of policy and a standardized process for tailoring. This solution will require that oversight and program management officials agree to early and continuous interactions and capturing of tailoring decisions in the acquisition strategy approved at Milestone B (or in the technology development strategy approved at Milestone A).[6] Both communities must also follow the tailored strategy afterward, lest deviations cause the entire set of tailoring decisions to be revisited.

---

[6]   A 2009 Government Accountability Office (GAO) report reaches a similar conclusion. See GAO, 2009b.

Acknowledgments

We would like to thank our sponsors—PEO (Ships) and AT&L/PSA/Naval Warfare—for their support and feedback throughout the study.

Special thanks are due to the many OSD and Navy officials who agreed to be interviewed as part of the study. Their participation, insight, and willingness to speak candidly about issues helped ensure that we were addressing real policy problems and greatly enhanced the usefulness of the research.

We also would like to thank our two reviewers: Irv Blickstein, a senior researcher at RAND, and Larrie Ferreiro, Director of Research, Defense Acquisition University. Their thoughtful comments on our draft report greatly improved the quality of the final product.

Any remaining errors are the sole responsibility of the authors.

Abbreviations

ACAT	acquisition category
AMRAAM	advanced medium-range air-to-air missile
AoA	Analysis of Alternatives
ARA	Acquisition Resources and Analysis
AT&L	Acquisition, Technology, and Logistics
BIW	Bath Iron Works
BY	base year
C4I	command, control, communications, computers, and intelligence
C4ISP	Command, Control, Communications, Computers, and Intelligence Support Plan
CAE	Component Acquisition Executive
CAIG	Cost Analysis Improvement Group
CDD	Capability Development Document
CDR	critical design review
CJCSI	Chairman of the Joint Chiefs of Staff Instruction
CMC	Commandant of the Marine Corps
CNO	Chief of Naval Operations
COEA	Cost and Operational Effectiveness Analysis
CONOPS	concept of operations
CPD	Capability Production Document
DAB	Defense Acquisition Board

DAE	Defense Acquisition Executive
DAMIR	Defense Acquisition Management Information Retrieval
DASN	Deputy Assistant Secretary of the Navy
DD&C	detail design and construction
DDG	guided missile destroyer
DFAR	Defense Federal Acquisition Regulation
DoD	Department of Defense
DoDD	Department of Defense Directive
DoDI	Department of Defense Instruction
DON	Department of the Navy
DR	decision review
DSARC	Defense Systems Acquisition Review Council
DT&E	developmental test and evaluation
EDM	Engineering Development Model
EMD	engineering and manufacturing development
EVM	earned value management
FAR	Federal Acquisition Regulation
FOC	full operational capability
FRP	full-rate production
FY	fiscal year
GAO	Government Accountability Office
GD	General Dynamics
GSA	General Services Administration
IBR	integrated baseline review
ICD	Initial Capabilities Document
ICE	Independent Cost Estimate
IOC	initial operational capability

IOT&E	initial operational test and evaluation
IPR	interim progress review
ISP	Information Support Plan
JCD	Joint Capabilities Document
JCIDS	Joint Capabilities Integration Development System
JHSV	joint high-speed vessel
JROC	Joint Requirements Oversight Council
KPP	key performance parameters
L&MR	Logistics and Materiel Readiness
LCS	littoral combat ship
LFT&E	live fire test and evaluation
LM	Lockheed Martin
LMSR	large, medium-speed, roll-on/roll-off (ships)
LRIP	low-rate initial production
MAIS	major automated information system
MDA	Milestone Decision Authority
MDAP	major defense acquisition program
MLP	mobile landing platform
MNS	Mission Needs Statement
MS	milestone
MSA	Materiel Solution Analysis
NASA	National Aeronautics and Space Administration
NASSCO	National Steel and Shipbuilding
NAVSEA	Naval Sea Systems Command
NGNN	Northrop Grumman Newport News
NGSS	Northrop Grumman Ship Systems
OIPT	Overarching Integrated Process Team

OPEVAL	operational evaluation
ORD	Operational Requirements Document
OSD	Office of the Secretary of Defense
OT&E	operational test and evaluation
OUSD (AT&L)	Office of the Under Secretary of Defense, Acquisition, Technology, and Logistics
PARM	Participating Acquisition Resource Manager
PAUC	program acquisition unit cost
PD	production and deployment
PDR	preliminary design review
PEO	Program Executive Office
PESHE	Programmatic Environment Safety and Occupational Health Evaluation
PI	program initiation
PM	Program Manager
PRR	production readiness review
PSA/NW	Portfolio Systems Acquisition, Naval Warfare
R&D	research and development
RDT&E	research, development, test, and evaluation
RFP	Request for Proposal
SAE	Service Acquisition Executive
SAMP	Single Acquisition Management Plan
SAR	Selected Acquisition Report
S&T	science and technology
SBIRS	space-based infrared system
SC	surface combatant
SCN	shipbuilding and conversion, Navy
SDD	system development and demonstration

SDR	system design review
SDS	System Design Specification
SECNAV	Secretary of the Navy
SECNAVINST	Secretary of the Navy Instruction
SECNAVNOTE	Secretary of the Navy Note
SEP	Systems Engineering Plan
SRR	system readiness review
SSE	Systems and Software Engineering
STAR	System Threat Assessment Report
T&E	test and evaluation
TD	technology development
TDS	Technology Development Strategy
TEMP	Test and Evaluation Master Plan
TRA	Technology Readiness Assessment
TRL	technology readiness level
USC	U.S. Code
USD (AT&L)	Under Secretary of Defense (Acquisition, Technology, and Logistics)

CHAPTER ONE

Introduction

Background

The management of a major defense acquisition program is an exceedingly complex challenge, involving the balancing and reconciling of many diverse interests stemming from different perspectives and constituencies. Program Managers might focus on near-term goals, including getting the new capability into the hands of the operational users as soon as possible. Others might place greater emphasis on minimizing risk (from their perspective) by insisting on extensive testing before production starts. Still others are concerned with ensuring that all funds are expended in ways fully consistent with the laws and the apparent intent of the Congress. These interests, perspectives, and objectives may conflict with each other and need to be resolved to move the program forward.

The U.S. Department of Defense (DoD) has a well-established set of policies, procedures, and organizations for acquisition program management and oversight, described in the 5000-series of directives and instructions.[1] These documents seek to identify and set guidelines to manage an acquisition path acceptable to all parties. The fact that there is a single overall process, applied to a wide variety of programs, has important consequences. One is that the policies and procedures, and the organization created to implement them, inevitably reflect certain characteristics and patterns that are inherent in most acquisition programs but might not be appropriate for some. For example, most systems go through three distinct stages: (a) establishing the need for a new item, (b) testing and validating the design through a technical development process, then (c) producing the item, often in large quantity. Frequently, the total cost of serial production will be many times the cost of developing, testing, and validating the design, thus justifying considerable time and effort for testing and validating a design.

Both internal stakeholders and external observers have criticized the acquisition management process for several decades, with numerous studies and reviews

---

[1]   Department of Defense, Under Secretary of Defense, Acquisition, Technology, and Logistics (AT&L), Department of Defense Instruction (DoDI) 5000.02, 2008; Deputy Secretary of Defense, Department of Defense Directive (DoDD) 5000.1, 2003.

seeking to "improve" it.[2] There has been occasional and incremental fine-tuning, and major new regulatory constraints have also been imposed, such as congressionally mandated procedures for operational testing (including live fire testing) before start of production, and, more recently, certification requirements at Milestones A and B.

Yet, overall, the process has proved remarkably resilient. The statutes and regulations that define acquisition processes reflect lessons from past experience and attempt to capture "best practice" for program management and oversight. That does not mean that the process is perfect, or even good enough; it does mean that attempts to modify it should be conducted with caution.

Not all kinds of weapon-system programs will comfortably fit the generic acquisition process model defined and implied by policy and regulation. In fact, every program is unique in one or more important ways. For some kinds of systems, including certain spacecraft and ships, there are no separate development units; all units produced, including the initial production article, are expected to go into operational service. Other systems might require exceptional emphasis on rapid execution to meet a pressing operational need, even if that requires compressing some of the standard steps in the process. The formal acquisition process is intended to be sufficiently flexible to accommodate such differences. Yet exercising such flexibility requires considerable initiative by the various stakeholders, from Program Managers to oversight officials, to adapt the procedures to each particular situation. The resultant "tailoring" is inevitably imperfect, at least as perceived by some of the participants.

Research Motivation

Ship[3] acquisition programs are one category of programs that appear to have some difficulty conforming to the traditional acquisition process model defined by DoD 5000 regulations. Ship programs are complex and have high unit costs as well as a long design and construction period. Ships are produced at lower rates and quantities than other major defense acquisition programs (MDAPs).[4] They also vary by type, with nuclear aircraft carriers and submarines, surface combatants, amphibious assault vehicles, and specialized and auxiliary ships all having unique production issues.

---

[2]   See for example Muñoz, 2008, 2009; Sherman, 2009; and Government Accountability Office (GAO), 2009b.

[3]   By *ships*, we are generally referring to the broad range of naval combatants including submarines, surface combatants, amphibious ships, and auxiliary and support vessels.

[4]   MDAP is an acquisition program designated as such by the Milestone Decision Authority (MDA), or estimated to require an eventual total expenditure for research, development, test, and evaluation (RDT&E) of more than $365 million in fiscal year (FY) 2000 constant dollars or more than $2.190 billion in procurement in FY 2000 constant dollars.

Stakeholders in the ship acquisition community, within both the Navy and the Office of the Secretary of Defense (OSD), have become increasingly frustrated that many of the same acquisition strategy and program structure issues, such as when low-rate initial production (LRIP) is authorized, must be addressed repeatedly, with little promise of a more permanent and consensual resolution. Many feel that some clarifying language in the 5000 instructions could help reduce these repetitive debates. This research attempts to define the differences and ambiguities more precisely and suggest ways to reach a more enduring understanding among stakeholders.

Problem Definition and Objectives

The objective of this research is to identify potential improvements in the policies and procedures used by the DoD in providing acquisition oversight to major ship procurement programs and by the U.S. Navy in supporting that oversight process. We give primary attention to (a) identifying any aspects of major ship acquisition programs that appear to deviate from the basic conceptual model underlying the 5000-series management process and (b) suggesting changes in either DoD or Navy acquisition policies and procedures that could ameliorate the undesirable consequences and improve the efficiency of major ship acquisition programs. Our focus will be DoDI 5000.02 and Secretary of the Navy Instruction (SECNAVINST) 5000.2D —instructions that form the baseline acquisition process for naval ships.[5]

Two sets of problems need to be addressed. One concerns how ships are specifically treated in acquisition policy, both statutory and regulatory. This includes the explicit treatment of ship programs and often takes the discretionary form of ". . . ships may . . ." or the more definitive form of ". . . for ship programs, this means x . . ."

The second set of issues concerns the perceived mismatch of ship-specific processes with the generic language used in policy and its implementation, particularly between the basic acquisition process and manufacturing processes specific to ship programs. For instance, if the engineering and manufacturing process for ships is sufficiently different from that of other weapon-system types, then some decision milestones or technical reviews may be either inapplicable or substantially different in timing and scope.

---

[5] Department of Defense, Under Secretary of Defense, Acquisition, Technology, and Logistics, 2008a, and Secretary of the Navy, 2008, respectively.

Research Approach

This project examines whether ships differ sufficiently from the standard acquisition model to deserve special consideration in these processes. To do this, we performed several research tasks between March 2008 and January 2009.

Task 1: Review Current Acquisition Policy with Respect to Shipbuilding Programs. For this task, we reviewed the key DoD and Navy acquisition policy documents (DoDI 5000.02 and SECNAVINST 5000.2D) to better define the traditional acquisition process model and determine where ship programs are currently treated differently in the acquisition process. Further, we compared and contrasted these instructions to illuminate where there are ambiguities within the policy with respect to ship programs.[6]

Task 2: Interview Key Stakeholders in the Acquisition Process. An important element of this research was an extensive series of interviews with stakeholders in the ship acquisition community. This included current and former ship program officials, managers, Program Executive Offices (PEOs), and others in Naval Sea Systems Command (NAVSEA); Navy oversight officials; and OSD officials from AT&L, General Council Comptroller, and Program Analysis and Evaluation. These interviews focused on three main questions:

- How do ships differ from other MDAPs?
- What issues or problems arise from those differences in the context of implementing the 5000 process?
- What regulatory changes are required to facilitate the acquisition process for ships?

The responses to these questions provide a catalog of specific ways in which ships differ from typical systems and how those differences affect their movement through the standard acquisition oversight review processes and related institutional practices.

We did not interview industry stakeholders in this study, although we have become familiar with industry views from other research. This study was focused on government stakeholders within the Navy and OSD with program management or oversight roles. It is within this community that the frustration regarding consistent interpretation and implementation of acquisition policy motivating this study occurs. We believe that the breadth of interviews we conducted adequately covers the issues we address.

Task 3: Review the Acquisition History of Major Ship and Nonship Programs Subject to the Management and Oversight Process Defined in DoDI 5000. We compared ship acquisition programs with other MDAPs that pass through

---

[6]   During the course of this research, the main, 2003, DoD acquisition regulation, DoDI 5000.2, was updated and reissued in 2008. Our analysis focuses on the most recent version of the regulation.

the traditional regulatory process. Specifically, we compared ship programs with aircraft, missiles, spacecraft, and ground vehicles. As we shall see in later chapters, the absence of separate RDT&E items dedicated exclusively to testing and relatively small production quantities is one way in which some ship programs differ from many acquisition programs.

We also explored how, and to what extent, these differences have affected how each system moves through the regulatory process. We prepared a database of programmatic information on recent acquisition programs, ships, and other system types to identify (1) the overall program structure, paying particular attention to the major decision points that occurred during the evolution of the program, and (2) how each program interacted with the regulatory oversight process (Defense Acquisition Board [DAB] review, etc.).

Task 4: Develop Suggestions for Improvement. We identified (and will review) special or unique aspects of ship programs that appear to require clarification or be not effectively managed in the standard acquisition oversight process, suggesting potential remedies or changes. These suggested improvements include possible changes to the 5000-series regulations and procedures that would explicitly provide for the special features of ship programs. We focus our suggestions on areas where the acquisition regulatory process is unlikely to accommodate the unique characteristics of ships and, therefore, is likely to impose costs on major ship acquisition programs.

Organization of This Monograph

The remainder of this monograph presents our analysis of the issues raised above. Each chapter roughly follows the outline of the tasks just described. Chapter Two examines how ship programs are treated in statute and regulation, with special attention to specific DoD and Navy directives and instructions. It identifies how ship programs are treated explicitly in regulations and begins to address the more nuanced mismatch between ship program characteristics and the traditional acquisition process. Chapter Three summarizes our interviews with stakeholders, presenting a wide range of opinions but also a set of themes common to many. Chapter Four examines ship and non-ship programs to identify differences in program structure and acquisition processes, validating some of what we heard from the stakeholders. Chapter Five explores how a ship program's management and oversight activities could best be structured in a hypothetical, unconstrained setting, with reference to how some key regulatory and process constraints affect ship acquisition programs. Chapter Six summarizes our findings and recommends actions that the Navy and OSD stakeholders can take to mitigate the problems we have identified.

CHAPTER TWO

How are Shipbuilding Acquisitions Treated Differently in Policy?

In this chapter, we summarize various policies, instructions, and guidebooks on the DoD acquisition process and how those policies differ for shipbuilding. We primarily focus on DoDI 5000.02 and SECNAVINST 5000.2D. We review each document for specific cases regarding shipbuilding programs and highlight relevant differences. As we will see, shipbuilding program initiation may begin earlier (e.g., at Milestone A) than in other programs, and production may be approved earlier (e.g., at Milestone B), meaning that reporting requirements may occur earlier as well. Although tailoring of processes can help accommodate these differences, it is not always consistent, leading to some ambiguities in implementation.

DoDI 5000.02 (2008)

Process and Definitions Differences

DoDI 5000.02 describes the operation and procedures of the department's acquisition system. This goal of the instruction is said to establish

> . . . a simplified and flexible management framework for translating capability needs and technology opportunities, based on approved capability needs, into stable, affordable, and well-managed acquisition programs that include weapon systems, services, and automated information systems (AISs).

The instruction outlines a phase-decision point process where programs proceed to the next phase of development only after successfully completing a milestone review. The acquisition process has five distinct phases: Materiel Solution Analysis (MSA),[1] technology development (TD),[2] engineering and manufacturing development (EMD),[3]

---

[1] This phase assesses potential material solutions.

[2] During the technology development phase, technology risks are reduced and the maturity of the desired technologies is assessed.

[3] The purpose of EMD is to continue to develop system capability.

production and deployment (PD),[4] and operations and support. There are three milestone decision points. DoDD 5000.02 states

- Milestone A (MS A)—the decision point between the MSA and TD phases. This milestone typically occurs once the Analysis of Alternatives (AoA) is completed and a specific technical solution is proposed. The AoA evaluates the various conceptual alternatives' ability to meet the mission need. Full funding for the next phase must also be in place. At MS A, the MDA approves the preferred material solution, approves the preliminary technology development strategy (TDS), and prepares a certification memo as required by statute.[5]
- Milestone B (MS B)—the decision point between the TD and EMD phases. MS B typically marks the formal program initiation point. By this milestone, the preliminary design review (PDR)[6] has usually been completed and the relevant technologies and manufacturing approach have been matured and (hopefully) demonstrated. The cost and schedule baselines also have been determined. Again, the MDA must also prepare a certification memo, as required by statute.
- Milestone C (MS C)—the decision point between the EMD and PD phases. This milestone typically authorizes a program to begin production at a low rate. By this point in the program, engineering and technical development is complete and all testing and operational assessments have been successful.

The acquisition process begins with the materiel development decision, which is mandatory for all programs. Figure 2.1 depicts the process framework taken from the instruction. At its core, the process develops a technical solution (e.g., a platform or system) for a needed capability. The milestones are checkpoints to validate that the development work is sufficient to proceed. Each step to the next phase represents an increasing commitment in resources and activity by the department. The underlying theme of the process is "try before buy." That is, the acquisition process emphasizes the need to demonstrate that one can realize the desired capabilities within the agreed budget and time.

The acquisition process seeks to be flexible, giving the PM and the MDA the ability to "exercise discretion and prudent business judgment to structure a tailored, responsive, and innovative program" (DoDI 5000.02, 2008). The MDA authorizes both entry into the process and the ability to proceed to the next program phase. The MDA has the authority "to tailor the regulatory information requirements and

---

[4] Production and deployment occur when an operational capability that satisfies mission needs is achieved.

[5] See U.S. Code (USC), Title 10, §2366a.

[6] PDR establishes a design baseline, which determines the cost and capability of the program being developed. The Program Manager (PM) plans the PDR before MS B and puts together a PDR report for the MDA to review.

**Figure 2.1
DoDI 5000.02 Acquisition Framework**

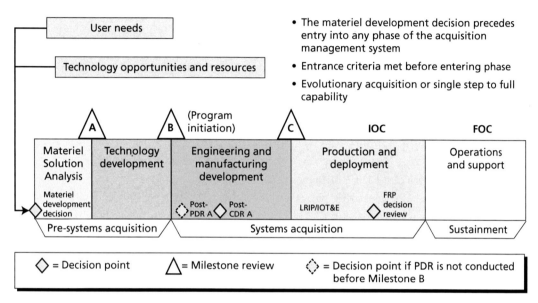

SOURCE: DoDI 5000.02.
RAND *MG991-2.1*

acquisition process procedures in this Instruction to achieve cost, schedule, and performance goals" (DoDI 5000.02, 2008).

One major difference for shipbuilding programs occurs at the start of the technology development phase (post–MS A). Typically during the TD phase, programs reduce technology risk by development activities and by identifying technical components to be integrated into the entire system. System development and program activities, such as procurement and engineering design, wait until the next phase. At the start of the EMD phase, the program is formally "initiated" (e.g., a formal program begins, a Program Manager is selected, a program office is set up, and major contracts may be let). However, shipbuilding programs can be initiated during technology development at MS A and therefore may begin their formal program activities earlier than other weapon systems.[7]

---

[7]   Specifically, "The MDA may initiate shipbuilding programs at the beginning of Technology Development. The information required by the tables in Enclosure 4 shall support program initiation. The CAIG [Cost Analysis Improvement Group] shall prepare a cost assessment in lieu of an Independent Cost Estimate (ICE), and the DoD Component shall provide a preliminary assessment of the maturity of key technologies. CAIG cost assessments for other acquisition category (ACAT) I and IA programs shall be prepared at the MDA's request" (DoDI 5000.02, 2008).

Similarly, the MS B decision for entrance into EMD is different for shipbuilding programs. MS B essentially authorizes the lead ship construction.[8] In practice, MS B authorizes the award of a detail design and lead ship construction contract (or the equivalent). In some cases, at the discretion of the MDA, only detail design is authorized, with exercise of the lead ship construction option in the contract authorized at a subsequent DAB-level program review. Therefore, ship programs can begin manufacture during the EMD phase, whereas in other programs, the technology must be sufficiently mature before production design or manufacturing is begun. This may mean that the technical activities, maturity, and information availability for ships may not match what is specified (or expected) in DoDI 5000.02.

The critical design review (CDR) is also a major milestone for ships (and other MDAPs). Successfully completing CDR indicates that the design is sufficiently mature to proceed into detail design for ships (or what might called "product design" more generally). DoDI 5000.02 (2008) places CDR after MS B and uses it as the demarcation decision point between the two aspects of EMD: The MDA conducts a "post-CDR assessment," successful completion of which ends integrated system design and approves entry into system capability and manufacturing process demonstration (see DoDI 5000.02, 2008, pp. 21–22).

For most programs, MS C authorizes entry into LRIP and the start of the production and deployment phase. But as the shipbuilding programs may start sooner and the production quantities (by number) tend to be lower for shipbuilding programs, the definition for LRIP changes for shipbuilding programs. Specifically, "LRIP for ships and satellites is production of items at the minimum quantity and rate that is feasible and that preserves the mobilization production base for that system" (DoDI 5000.02, 2008). Perhaps more important, LRIP for ships is functionally approved at MS B (or soon thereafter)—the lead ship can be considered the start of initial low-rate production.

Except for a unique definition of LRIP, DoDI 5000.02 does not have any specific language for ship programs with respect to MS C. This lack of language leads to several ambiguities. For example, whereas MS B may authorize the lead and some follow ships, there is no corresponding language that defines when LRIP occurs for ships if production begins at MS B. If LRIP begins with the first of class, then LRIP should logically move to MS B. Yet, DoDI 5000.02 does not specify that LRIP for ship programs may occur before MS C. Rather, it states, "Milestone C authorizes entry into LRIP (for MDAPs and major systems), into production or procurement (for non-major systems

---

[8]   More specifically, "For shipbuilding programs, the required program information shall be updated in support of the Milestone B decision, and the ICE shall be completed. The lead ship in a class shall normally be authorized at Milestone B. Technology Readiness Assessments shall consider the risk associated with critical subsystems prior to ship installation. Long lead for follow ships may be initially authorized at Milestone B, with final authorization and follow ship approval by the MDA dependent on completion of critical subsystem demonstration and an updated assessment of technology maturity" (DoDI 5000.02, 2008).

that do not require LRIP) or into limited deployment in support of operational testing for major automated information system (MAIS) programs or software-intensive systems with no production components" (DoDI 5000.02, 2008). Here, the definition of LRIP under MS C for ships is ambiguous, as the exception at MS B is not reinforced at MS C. As we shall see, several individuals in the acquisition community find the language around MS C for ships ambiguous, given the existing tailoring for ship programs of MS B.

Similarly, there is no specific language for ships regarding full-rate production (FRP). For many ship programs, the production rate does not change substantially through the life of the program. So if the production rates for LRIP and FRP are the same, what does the FRP decision authorize—continued production? Furthermore, per regulation, a full-rate production decision cannot be made until initial operational test and evaluation (IOT&E) and the beyond-LRIP report are complete. For many ship programs, a substantial fraction of total production will be authorized at that point (when testing is complete). Does an FRP decision in this case make sense? We will return to these issues in a later chapter.

Statutory, Regulatory, and Contract Reporting Information and Milestone Requirements

Because shipbuilding programs may begin with MS A, these programs may be required to generate reports sooner than most programs. Tables 2.1 and 2.2 highlight the earlier statutory and regulatory reporting requirements for shipbuilding programs that have program initiation at MS A.[9] Most of these reporting requirements must be updated at successive reviews and milestones. So, a shipbuilding program that starts the process earlier may have more update work to do than other programs. Perhaps more important, the level of detail at which these reporting requirements are normally treated will be different for a ship program documenting formal program initiation at MS A. In particular, the system design is more mature at MS B, when these information requirements are usually addressed. This sets up potential tension between the Navy and OSD stakeholders regarding the sufficiency of information available at MS A. This issue is not directly addressed in DoDI 5000.02. However, for all programs, "MDAs may tailor regulatory program information to fit the particular conditions of an individual program. Decisions to tailor regulatory information requirements shall be documented by the MDA" (DoDI 5000.02, 2008).

---

[9] If a general requirement exists to report at MS A, then we do not list the requirement in Tables 2.1 and 2.2. Any program would have to report, given the requirement to go through the MSA phase. There would be additional reporting requirements for ships if program initiation occurred at some time between MS A and MS B. For simplicity, we have not identified all these cases. Readers interested in details or in the content of each of these reporting requirements should refer to the *Defense Acquisition Guidebook* (Defense Acquisition University, 2006) for more information.

Table 2.1
Statutory Reporting Requirements

Information/Reporting Requirement	Shipbuilding Start	Typical Start
Clinger-Cohen Act compliance	Program initiation (PI)	MS B
Registration of mission-critical and mission-essential information systems	PI	MS B or MS C
Programmatic Environment Safety and Occupational Health Evaluation (PESHE)	PI	MS B
Selected Acquisition Report (SAR)	PI	MS B
Independent Cost Estimate	PI (cost assessment only)	MS B
Manpower estimate (reviewed by the Office of the Under Secretary of Defense for Personnel and Readiness)	PI	MS B
Acquisition program baseline	PI	MS B

SOURCE: DoDI 5000.02, 2008.

Table 2.2
Regulatory Reporting Requirements

Information/Reporting Requirement	Shipbuilding Start	Typical Start
Capability Development Document (CDD)	PI	MS B
Acquisition Strategy	PI	MS B
Analysis of Alternatives	PI	MS B
Technology Readiness Assessment	PI (preliminary assessment)	MS B
Command, Control, Communications, Computers, and Intelligence Support Plan (C4ISP)	PI	MS B
Component cost analysis	PI	MS B
Cost Analysis Requirements Description	PI	MS B
Exit criteria	PI	MS B
Information Support Plan (ISP)	PI	MS B
System Threat Assessment Report (STAR)	PI	MS B

SOURCE: DoDI 5000.02, 2008.

SECNAVINST 5000.2D (2008)

This instruction from the Office of the Secretary of the Navy (SECNAV) outlines the Department of the Navy's (DON's) specific implementation of the acquisition process. It largely mirrors DoDI 5000.02. Again, the system is stated to be flexible: "All MDAs should promote maximum flexibility in tailoring programs under their oversight." The current SECNAVINST *predates* the recent revisions to DoD 5000.02; the differences and similarities we highlight below might change when the SECNAVINST is revised.

SECNAVINST 5000.2D reaffirms that shipbuilding program initiation may begin with MS A: "Normally program initiation will occur at Milestone B, but may occur at the start of Technology Development, Milestone A, for shipbuilding programs. For shipbuilding programs not started at Milestone A, the CDD will be approved prior to the start of functional design." Also, "The MDA may approve program initiation for shipbuilding programs at Milestone A, the beginning of the Technology Development phase."

This earlier start allows ship design work to be concurrent with technology development.[10] This concurrent development and design, however, runs counter to the spirit of DoDI 5000.02, which seeks to reduce technical risk through prototyping in the technology development phase.

The SECNAVINST offers a slightly different interpretation of MS B than the DoDI for shipbuilding programs. Specifically, the SECNAVINST views MS B as the approval for the lead and initial follow ships.[11] The view that LRIP for ships is the minimum sustaining rate implies that MS B is also an LRIP decision for ships (by SECNAVINST 5000.2D). The SECNAVINST also moves the combat system technology demonstration before the ship "installation" point (i.e., testing of the system must occur before its placement on the ship).

Unlike DoDI 5000.02, the SECNAVINST specifically tailors MS C and the FRP DR for ships. It introduces the possibility of combining MS C and the FRP DR into a single event.[12] So what does the FRP DR mean for ships? The instruction notes,

---

[10] As the instruction notes, "Technology development is normally part of pre-systems acquisition effort conducted prior to program initiation. Technology to be used in the initial and subsequent increments of a program shall have been demonstrated in a relevant environment. Shipbuilding programs may be initiated at Milestone A in order to start Ship Design concurrent with sub-system/component technology development."

[11] The instruction states, "In the case of shipbuilding, lead and initial follow ships are normally approved at Milestone B. The follow ships that are approved at Milestone B shall be sufficient quantities to maintain shipyard construction continuity until the FRP decision review (DR). Critical sub-systems such as combat systems shall be demonstrated prior to lead and follow ship installation as directed by the MDA given the level of technology maturity and the associated risk."

[12] Specifically, the instruction notes, "For those programs that do not require LRIP and have completed required IOT&E or for shipbuilding programs where follow ships are initially approved at Milestone B, Milestone C and the FRP DR may be combined into a single program decision point as long as all of the required program information for both Milestone C and FRP DR are satisfied."

"In the case of shipbuilding programs, the FRP DR shall be held to provide the MDA the results of the completion of IOT&E, authorize the construction of the remaining follow ships." The FRP decision for ships is the acceptance of the test results to continue production. It is not necessarily a production-rate decision.

This SECNAVINST formalizes an internal-to-the-Navy acquisition process, referred to as the "two-pass/six-gate" acquisition process.[13] The intent of this process is to prepare the Navy to better support the DoDI 5000.02 process. The notice establishes enhanced oversight focused on the requirements-generation process and cost-risk evaluation. The system adds six additional milestones to the Navy internal acquisition process. These new milestones are grouped into two "passes." Pass 1 comprises the three requirements gates, including the concept refinement phase. This phase is led by the Chief of Naval Operations (CNO)/Commandant of the Marine Corps (CMC). Pass 1 begins before Materiel Solution Analysis and ends with CDD approval just before to MS A.[14] Pass 2 starts at the completion of Pass 1 just before MS A and ends at program completion. The notice specifically says that naval nuclear propulsion remains the domain of NAVSEA08.

Pass 1: Concept Decision and Concept Refinement Phase

- Gate 1: Culminates in the Navy's approval of the Initial Capabilities Document (ICD) to submit to J-8. This gate also validates the AoA plan and approves the start of concept decision.
- Gate 2: Occurs after the completion of the AoA but before MS A. It reviews the AoA content, approves the Service's preferred alternatives, approves the start of the CDD and concept of operations (CONOPS) generation, and proceeds to the next step. If program initiation is MS A, then the next step is Gate 3. If not, the next event is MS A.
- Gate 3: Reviews the preliminary design and cost outputs. Specifically, it reviews the DON-generated CONOPS and CDD; validates the System Design Specification (SDS) development plan; and reviews cost, risk, and budget.

Pass 2: Technology Development Phase

- Gate 4: Approves the SDS and proceeds to MS B or Gate 5.

---

[13] Previously, SECNAV Notice (SECNAVNOTE) 5000 (Secretary of the Navy, 2008a). Another good reference for this new system is Department of the Navy, March 2008a.

[14] For a program that is initiated at MS A. If program initiation is MS B, then the CDD is approved soon after MS A.

Pass 2, Continued: System Development and Demonstration Phase[15]

- Gate 5: Checks that all items are complete before releasing the system development and demonstration (SDD) or a Request for Proposal (RFP). It may be combined with MS B.
- Gate 6: Evaluates the readiness for production and the sufficiency of the earned value management (EVM) system program baseline and integrated baseline review (IBR). It follows the award of the SDD contract and IBR.

The process differs slightly by when program initiation occurs. For ship programs that have initiation as MS A, the process is shown in Figure 2.2.

Other Relevant Acquisition Documents

Several other acquisition-related documents pertain to shipbuilding programs, including

- Chairman of the Joint Chiefs of Staff Instruction (CJCSI): CJCSI 3170.01F (May 1, 2007a)
- Chairman of the Joint Chiefs of Staff Manual: CJCSI 3170.01C (May 1, 2007b)
- *Acquisition and Capabilities Guidebook:* Department of the Navy (undated)
- *Defense Acquisition Guidebook:* Defense Acquisition University (November 2006)
- *Federal Acquisition Regulation (FAR)*: General Services Administration (GSA), DoD, and NASA (March 2005)
- Defense Federal Acquisition Regulation (DFAR): Department of Defense (April 23, 2008).

We discuss in detail in Appendix A how these documents address shipbuilding. The first two documents have minimal tailoring for ship programs. The second two documents are guidance, are not mandatory, and do not provide guidance regarding the acquisition (5000) process itself; rather, the differences for ships focus on testing. The last two documents are mandatory acquisition regulations that note some specific differences for ships. Many of the differences are in the implementation details and in how certain program activities (e.g. design authority, contracting and sourcing requirements, change orders) can proceed. None specifically address the 5000 process.

---

[15] This is one example where DoDI 5000.02 and SECNAVINST 5000.2D are out of synchronization. The description of the phases uses the prior DoD 5000 terminology, where the EMD phase was called System Development and Demonstration.

Figure 2.2
Two-Pass/Six-Gate Process for Program Initiation at MS A

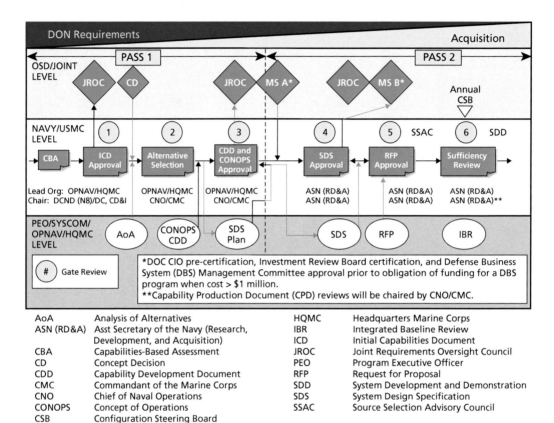

AoA	Analysis of Alternatives
ASN (RD&A)	Asst Secretary of the Navy (Research, Development, and Acquisition)
CBA	Capabilities-Based Assessment
CD	Concept Decision
CDD	Capability Development Document
CMC	Commandant of the Marine Corps
CNO	Chief of Naval Operations
CONOPS	Concept of Operations
CSB	Configuration Steering Board

HQMC	Headquarters Marine Corps
IBR	Integrated Baseline Review
ICD	Initial Capabilities Document
JROC	Joint Requirements Oversight Council
PEO	Program Executive Officer
RFP	Request for Proposal
SDD	System Development and Demonstration
SDS	System Design Specification
SSAC	Source Selection Advisory Council

SOURCE: SECNAVINST 5000.2D.
RAND MG991-2.2

Discussion of DoDI 5000.02 and SECNAVINST 5000.2D with Respect to Shipbuilding

Although both DoDI 5000.02 and SECNAVINST 5000.2D affirm that the acquisition process should be easily tailored and flexible, they also have a few inconsistencies, which may lead to confusion (as the interviews we discuss in the next chapter also confirm). These differences suggest areas where both documents could be further clarified to make them consistent.

One major difference between the documents is in the authorization of lead and follow ships at MS B. These differences are subtle. The SECNAVINST indicates that the lead *and* follow ships are approved at MS B. DoDI 5000.02 indicates that only the first-of-class ship and long-lead for follow ships are authorized at MS B; it formally approves follow ships when subsystems are demonstrated and the technical maturity is "updated." This difference between the two documents highlights possible ambiguity on how many ships are or can be authorized at MS B.

This difference on what is authorized at MS B may also confuse what constitutes the start of LRIP for ship programs. Both documents are consistent in their definition of what constitutes LRIP for ship programs, but DoDI 5000.02 does not clearly state that MS B is the start of LRIP. Does the LRIP DR occur when the MDA approves the follow ships? The SECNAVINST indirectly implies that MS B is also the LRIP decision for ships (as it uses language that is consistent with the definition of LRIP).

With respect to MS C and FRP DR, DoDI 5000.02 does not have language for tailoring these decision points. It leaves open to multiple interpretations what constitutes FRP and what the MS C decision constitutes for ships. The SECNAVINST talks about combining the FRP DR and MS C into one review that serves as a continuing production decision for the MDA.

We shall see in the next chapter that major stakeholders in the acquisition process for ships are confused over these differences and ambiguities.

CHAPTER THREE

Stakeholder Interviews

In the previous chapter, we noted how differences and ambiguities in the acquisition instructions can lead to uncertain interpretations of policy for ships. To help determine whether these uncertainties do, in fact, exist in practice, we interviewed a number of stakeholders involved in the acquisition of ships. This chapter summarizes these interviews.

The interviews focused broadly on acquisition issues related to ships. We covered such topics as perspectives on the differences relevant to acquisition policy for ships and other MDAPs, areas where problems exist in policy for ships, and areas where acquisition policy could be improved for ships. In this chapter, we report only the responses from those interviews and do not attempt to evaluate their validity.[1]

These interviews confirmed that ships are perceived to differ from other MDAPs in ways that affect the acquisition process and that certain areas of policy are ambiguous. We also found varying opinions on how acquisition policy for ships could be improved.

Who We Interviewed

RAND analysts conducted a total of 25 interviews with over 30 individuals. We selected interviewees to broadly represent the acquisition community, including both OSD and NAVSEA (including the project management community).

From OSD, we interviewed personnel from

- Office of the Under Secretary of Defense (OUSD/AT&L)
  - Acquisition Resources and Analysis (ARA)
  - Systems and Software Engineering (SSE)

---

[1] We attempted to achieve consistency of interviews through a question template, but not all interviewees commented on all issues addressed in this monograph. Therefore, reporting the number of interviewees who stated certain facts would be misleading. Lack of comment does not imply a perceived lack of importance or significance of the issue being discussed.

 - Logistics and Materiel Readiness (L&MR)
 - Defense Portfolio Systems Acquisition, Naval Warfare (PSA/NW)
- CAIG
- Director, Operational Test and Evaluation
- Office of the General Counsel (DoD)
- Office of the Under Secretary of Defense, Comptroller's Office.

From the Navy, we interviewed personnel from

- Deputy Assistant Secretary of the Navy (DASN) ships
- NAVSEA
- PEO ships
- PEO submarines
- Programs
  - CVN-78
  - Joint high-speed vessel (JHSV)
  - Littoral combat ship (LCS)
  - CG(X)
  - Large, medium-speed, roll-on/roll-off (LMSR) ships and mobile landing platform (MLP)
  - Amphibious ships (e.g., LPD-17)
  - DDG-1000
- Chief of Naval Engineering (NAVSEA 05d).

What the Interviews Covered

The interviews covered a range of topics and three broad research questions. Each interviewee was asked

- How are ships different from other MDAPs?
- What issues or problems arise from those differences in the context of implementing the 5000 process?
- What regulatory changes are required to facilitate the acquisition process for ships?

Although the interviewees participate in a wide variety of activities, their answers to these questions were often similar. We first summarize the common themes for each question and then highlight other important considerations.

What Is Different About Ships?

All interviewees, except one individual from OUSD (AT&L), agreed that ships differ from other major weapons systems that the DoD procures. The following list summarizes the most commonly identified differences.

- length of time to design and build
- influence of industrial/political factors
- concurrency of design and build
- higher complexity
- low quantity/production rate
- high unit cost
- type of funding
- test and evaluation (T&E) approaches.

The amount of time it takes to design and produce a ship was the primary difference noted by interviewees from both DoD and NAVSEA. Ship size and complexity are largely responsible for the long design and construction time. Ships, particularly nuclear vessels, can require long lead items and must be ordered well in advance of production—often far earlier than other MDAPs.

The second most commonly identified difference was the role of the industrial base and political factors in influencing the acquisition process. Many interviewees remarked that the role of the industrial base and political factors are more prominent for ships than for other major weapon-system programs. Interviewees noted in particular the role of the industrial base in the development of acquisition strategies and in influencing specific acquisition decisions.

OSD and Navy interviewees also noted that the high level of concurrency between the design and build process was significantly different for ships. Submarine and aircraft carrier builders have adopted a more seamless and integrated design and build process, referred to as the Integrated Product and Process Development model.

The development and production of different systems and sections of the ship also occur at different times. Interviewees noted that this concurrency is necessary to shorten long design and development time lines. However, some noted that the overlap of research and development (R&D) efforts for planned ship systems and the design of the ship can cause problems.

The number of ship systems designed, produced, and integrated contributes to ship complexity, another difference interviewees identified between ships and other MDAPs. Many interviewees noted that ships are a "system of systems." A ship program can have a number of Participating Acquisition Resource Managers (PARMs),[2] which

---

[2] A PARM is a resource manager that is typically responsible for providing some service, system, or component to the program.

need to be managed and integrated. One ship program had more than 70 PARMs. Unlike most MDAPs, many ship MDAPs include other MDAPs as part of the ship program. One interviewee noted that ships are more complex because they have more redundant systems than other platforms.

Interviewees noted that ships were procured in lower quantities and for larger unit prices than are other major weapon systems. In only a few instances, ships are procured in larger quantities but at low production rates (one or two per year) and are still costly. Interviewees offered few examples of low-cost ships or ships with high quantities or production rates. Interviewees noted that because of relatively high unit cost and low total production quantities, ship programs do not typically design and build prototype units designated solely for test. The first unit produced is deployable and tasked for service. These characteristics are shared by some satellite programs.

Interviewees also noted that funding for ships differs from that for other MDAPs. Ships (including the first of class) are typically funded within a unique appropriation: shipbuilding and conversion, Navy (SCN), although they have also been funded using RDT&E (LCS) and the National Defense Sealift Fund appropriation (T-AKE, JHSV). The SCN funding for the lead ship typically includes detail design and final development activities—activities that in other programs use RDT&E funds. Many interviewees discussed how the first-of-class of other weapon systems are typically appropriated with RDT&E. The SCN appropriations are typically for a period of five to seven years. The rules governing use of, and accounting for, SCN funds are closer to those for procurement funding than to those for more flexible RDT&E appropriations. Interviewees also discussed other unique aspects of funding ships. Ships have a "Prior Year Completion" line item to fix cost shortfalls; no other system has such a line item. It reflects, in part, the long process of building a ship and the difficulty in estimating the cost. Unlike many other weapon systems, ships generally require full funding in the year appropriated, but Congress can approve split funding, where the cost of the ship is split between two years. This has been done for carriers and large deck amphibious ships where the single unit acquisition cost is very large.[3]

Nearly every interviewee noted testing and evaluation as another area where ships differ from other MDAPs, although views varied between OSD and the project management community. Many individuals in OSD noted that the testing requirements and production decision points were mismatched for ships. For instance, many ships are on contract before testing is complete. The project management community felt that shipbuilding did not align well with the testing requirements, because the level of testing or detail of documentation required was excessive. The PMs also questioned the value of much of the testing. Many felt that the testing requirements offered little and cost a lot. The testing community felt that useful testing could be performed for ships.

---

[3] The DDG 1000's dual lead ships were also split-funded, which required special approval by Congress.

Many interviewees noted cultural differences between the ship community and the acquisition community. Each uses different terminology and language. Interviewees noted that the internal process and culture of the Navy may not mesh with DoDI 5000. For example, "delivery" of a product means something different for ships than for other communities. The ship completes construction and then undergoes a period of evaluation by the user. The ship is "accepted" into the fleet after this evaluation period. This period of evaluation is typically referred to as a post-delivery and outfitting period, where any problems are fixed and systems are upgraded.

Some other differences interviewees noted for ships were

- Ships can enter the acquisition process at MS A.
- Ship programs cannot easily "de-scope" or reduce their quantity.
- Ships are more like a major military construction project than weapon-system procurement for many reasons, including the similarity in the way they are constructed and the habitability requirements.

Although nearly all interviewees agreed that ships are different from other major defense acquisition programs, opinions varied regarding the implications of these differences for the acquisition process. Some interviewees felt that the current ability to tailor the process sufficiently addressed the unique aspects of ships. Others felt that tailoring was either too cumbersome or inadequately addressed the unique aspects of a ship program.

How Is Implementing the DoDI 5000 Process Difficult for Ships?

The RAND team received a wide variety of comments regarding ships and the DoD acquisition process. Some interviewees noted that differences between ships and other major weapon systems—including quantities, the amount of time to build, and the overlap between the design and construction phases—made it more difficult for ships to fit into the DoDI 5000 process. Others noted challenges not necessarily caused by such differences. Some suggested that there are no unique challenges for ships but that much of the process is unnecessary or serves no purpose for ships.

Interviewees commented that the phasing and degree to which requirements are met is different for ships. Some claimed that process tailoring is sufficient to address the unique phasing and requirements, whereas others felt that the tailoring was insufficient. One interviewee noted that all weapon-system acquisition programs are the same—"they all need tailoring." In other cases, ambiguities in language or the interpretation of DoDI 5000 was thought to make implementation of the 5000 process more difficult for ships. Areas that interviewees identified as distinct challenges for ships included

- interpretation issues in DoDI 5000
  - MS B/start of production/LRIP
  - FRP/MS C
- content and timing of documentation requirements
- test and evaluation issues
- statutory issues
- other DoDI 5000 policy/process issues.

DoDI 5000 Interpretation Issues

Both the OSD and Navy program management community felt that issues with semantics and the interpretation of DoDI 5000 have led to a need to address the same issues repeatedly. Some interviewees felt that there is too much variability in the interpretation of the instruction. OSD oversight officials interpreting regulatory requirements differently may ask for different things. One interviewee noted in particular the high variability in the content of acquisition strategies and lack of a template for them. The noted consequence of this is the reinvention of statutory and regulatory information requirements for each program. Many noted that the ability to tailor the process made it unclear what the "'requirements' requirements" actually are. Others said that the terminology used in shipbuilding was unique, causing some confusion. For example, the definition of a "delivery" or "production representative unit" is not the same as in nonship program definitions. Specifically noted as being challenging to interpret or define for ships were

- what MS B authorizes for ships
  - meaning of LRIP
  - definition of start of production
- meaning of MS C
- definition of FRP.

Some interviewees commented that the meaning of MS B was ambiguous for ships. They claimed that the definition of program initiation, which DoDI 5000 defines as occurring at MS B, was unclear. By MS B, a ship program office has been established, significant resources have gone into the program, and one or more design contracts have been awarded. This, some interviewees argued, means that program initiation (establishment of the program office) occurs long before MS B. Interviewees also noted that MS C typically authorizes the start of production and LRIP. For ships, however, lead ship and long-lead material are typically authorized at MS B. Interviewees pointed out that lead ship and long-lead material authorization is functionally the start of production.

The interpretation of LRIP (for ships) was unclear to most of the stakeholders we interviewed. The most commonly cited reason for this ambiguity was the fact that many ships can be on contract before authorization of LRIP, which formally occurs at MS C. Because the total number of ships on contract before MS C can be a significant percentage of the total class, interviewees said the LRIP decision at MS C was ambiguous. If additional ships are to be procured, the procurement rate is not likely to change. Some interviewees questioned the usefulness of defining LRIP at all in cases where only a single ship will be built, the production rate does not change (i.e., one per year), or where the classes are very small.

Interviewees noted that MS C was also ambiguous for ships. They pointed out that "production" of some number of ships may have commenced before the "production decision" of MS C. The number of ships that would constitute LRIP may also have already been contracted. In these cases, the meaning of MS C is questionable. Still, legislative requirements for MS C must be satisfied for all programs, including ships. Because the new acquisition legislation allows a program baseline to be changed only at a major milestone, some interviewees felt that MS C should be maintained to facilitate baseline revisions.

For some ship classes, a full-rate production decision may never be required. The rate of production will not be increased; there may be no additional production at all. In such cases, interviewees said that there was no point to having an FRP decision.

Content and Timing of Documentation Requirements

The issue most commonly raised by the project management community was the content and timing of the currently required documentation. Interviewees specifically questioned the utility of much of the documentation required. The PMs felt that the level of detail required in many of the documents was too great yet was still increasing. Many PMs noted that the level of detail desired upfront exceeded the information then available. Documents requiring too much detail, as identified by PMs, were the Test and Evaluation Master Plan (TEMP), the System Engineering Plan, and the Acquisition Strategy.

Many interviewees described the challenge associated with defining a single measure of design maturity for the critical design review and preliminary design review. Although some programs did not have these reviews, interviewees from programs that did felt that they were particularly challenging. For example, we discussed above how ship programs comprise many systems, each with its own technical maturity. Other programs, in fact, may manage some of these systems. So defining a single maturity value for a system-of-systems might be difficult or misleading.

OSD and Navy program management interviewees also noted that the timing of document production could be problematic. They said that the Capability Development Document, defining the capabilities that ship designers need to consider, was too late at MS B and should be created before the Systems Engineering Plan (SEP).[4] Beginning production of ships at or shortly after MS B does not leave much time for CDD capability definitions to mature or ensure that these capabilities have been designed into the ship. Similarly, interviewees said that having the Technology Readiness Assessment (TRA) at MS B is also too late. The imminent production of the lead ship leaves little time to adjust the design before production, should a technology not be ready. At the same time, project management interviewees said that achieving a technology readiness level[5] (TRL) of 6 by MS B for all ship systems was very challenging. The design and development of systems are staggered, with some being very mature at MS B, with a TRL of at least 6, and others still in development. In addition, several interviewees noted that two documents in the process, the AoA and the SEP, add very little value.

The content and timing of documentation requirements was less an issue for OSD interviewees. This was expected, given that OSD is the main consumer of the information. Nevertheless, interviewees noted that the timing of test and evaluation activities was not well aligned with milestones. They also noted redundancy between the Acquisition Strategy and the Technology Development Strategy.[6]

Test and Evaluation

Another major issue identified by interviewees was test and evaluation for ships. DoD officials felt that the timing of test and evaluation activities, relative to milestones, was more challenging for ships than for other MDAPs. Industrial base considerations typically lead to a number of ships being contracted before completion of testing. Yet, interviewees noted, increasing the amount of time between the authorization of lead ship and follow ships to allow for more testing can create problems for the industrial base. Some in DoD said that incremental testing could help, but others did not feel it was an acceptable substitution for integrated, system testing.

---

[4] "The SEP describes the program's overall technical approach, including systems engineering processes; resources; and key technical tasks, activities, and events along with their metrics and success criteria. Integration or linkage with other program management control efforts, such as integrated master plans, integrated master schedules, technical performance measures, risk management, and earned value management, is fundamental to successful application" (Defense Acquisition University, 2006).

[5] Technology readiness level is a rating of technology maturity. Ratings go from 1 to 9 with 1 being the least technically mature. Chapter 10 of the *Defense Acquisition Guidebook* (Defense Acquisition University, 2006) states, "The DoD Component Science and Technology (S&T) Executive directs the Technology Readiness Assessment and, for ACAT ID and ACAT IAM programs, submits the findings to the Component Acquisition Executive (CAE) who should submit his or her report to the DUSD(S&T) with a recommended technology readiness level (TRL) (or some equivalent assessment) for each critical technology."

[6] The Acquisition Strategy is required at MS B and the Technology Development Strategy is required at MS A. There is some redundancy in content, since the two documents serve essentially the same purpose: describing the program execution plan across a range of functional activities. See Chapter 2 of the *Defense Acquisition Guidebook* (Defense Acquisition University, 2006).

The PM community felt that there was not enough flexibility or tailoring allowed in the testing of ships. Although other MDAPs can quickly produce prototype systems for testing, ships cannot. In some instances, subsystem prototypes have been developed and used to successfully satisfy testing requirements. In other instances, PMs have complained that much of the early testing performed was not given due credit toward meeting testing requirements. Part of the issue here is the definition of an acceptable prototype. Many interviewees said that they did not know what an acceptable prototype was for ships, specifically, whether subcomponent or subsystem prototyping was sufficient to meet system prototyping requirements. Some interviewees pointed out that prototyping and other testing mechanisms used to accomplish IOT&E activities in the 5000 process are more costly for ships. They suggested that to cut costs, retired ships could be used for shock and other testing of subsystems before the new ship is built. In other words, assuming that the retired ships are representative of the new ship, test results from subsystem performance when installed on the retired ship could be used as a proxy for performance when installed on the new ship. Although there is currently a U.S. Navy test ship for combat-systems testing, the interviewees said that other attempts to establish surrogate testing platforms had not received adequate funding. The interviewees also felt that live fire testing for ships was unnecessary and far too costly .

Finally, IOT&E personnel said that test and evaluation is treated differently for different classes of ships, with carriers, submarines, and surface ships evaluated differently. This comment illustrates how the acquisition process may differ even from ship to ship.

Statutory Requirements

Few interviewees pointed out statutory requirements that are difficult to meet. Most were concerned with new statutory requirements that may be difficult for ship programs to satisfy. The Fiscal Year 2008 Defense Authorization Act, §943, revising Title 10 USC 2366b, requires certifications at MS A that many of our interviewees expected to be difficult for ships to satisfy. They said that the items now required at MS A are not typically available for ships by that point in the process.[7]

Other 5000 Policy and Process Issues

Interviewees identified a number of other issues for ships within the current 5000 policy and process. These included

- Changes to policy leading to unclear requirements. There is currently no feedback loop for PMs or others to identify how changes in policy have affected the ability

---

[7] The certifications include a complete cost estimate, a completed AoA, and other items related to program planning. See also USD (AT&L), 2009.

to successfully manage and complete a program. One such example provided was that of the policy on prototyping. Many individuals were unclear as to what was an acceptable prototype.

- Current processes not accommodating a common problem for ships: technological obsolescence and configuration management. Many PMs noted that the obsolescence problem is a result of the long time it takes to design and construct a ship. It is also difficult to specify technology refresh in the current acquisition process. As noted, ships are a system of systems, a fact not handled well by the current process.
- A 5000 model that, in the view of PM interviewees, is too serial. At the same time, some in DoD felt that there is too much overlap between design and production for ships. The two communities clearly have conflicting views here.
- There is relatively little program oversight required beyond MS C.

How Should the DoD 5000 Process Be Changed?

Most interviewees did not think the 5000 process was "broken," but many suggested ways to improve it to better accommodate unique ship challenges.

Currently, the 5000 process is said to be flexible and can be tailored to meet the unique needs of each individual MDAP. Yet several interviewees said that they did very little tailoring or that tailoring was too difficult. One PM stated that any tailoring resulted in more reviews and meetings, creating little incentive to seek it. Many suggested improving the ability to tailor the acquisition process but without formalizing tailoring. This could be done through improved guidance, setting of expectations, and standardization of some requirements. Interviewees also felt that cultural changes would be required to ensure that tailoring is acceptable and recognized as essential to ship programs. This could be accomplished by documenting those areas of the 5000 process where tailoring is allowed, including a description of acceptable tailoring. Interviewees suggested outlining in the Acquisition Strategy how a program is tailored.

Many interviewees felt that the meaning of LRIP, MS B, FRP, and MS C should be rethought and redefined accordingly. Some felt that LRIP and FRP should not apply to ships at all. Others felt that a clarification of the language in DoDI 5000 would reduce current confusion, while allowing the LRIP and FRP decisions to be made for programs where it makes sense to do so. Some felt that MS B and MS C should be combined for ships, because there is typically no difference between initial production and FRP, and production decisions are typically made at MS B, not at MS C. Others suggested a realignment or redefinition of the phase between MS B and MS C to focus more on technology risk reduction, which is now supposed to occur between MS A and MS B.

Some interviewees suggested revising the testing requirements and approach for ships. DoD interviewees suggested more and earlier component- or subsystem-level testing for ships. They suggested that this could be done through increased use of prototyping and adoption of more surrogate testing platforms. The program management community suggested streamlining the TDS and SEP to an appropriate level of detail, recognizing that ships are a collection of systems. The PM community also suggested that the system engineering process should be simplified and that minimal acceptable live fire testing requirements should be based on survivability.

Some, but not all, interviewees felt that an addendum or annex to DoDI 5000.02 would offer more specific guidance for resolving many of these issues. The annex would describe how ships can fit into the 5000 process. For example, the annex or addendum could clarify when a production decision occurs for ships and what that production decision entails. It could cover how to address the identified testing issues and what to include in the T&E strategy. It could also describe or outline an acceptable level of detail for documents or specify a rule set for determining an acceptable level of detail.

The PM community suggested checks and balances to mitigate the tension between the desire of OSD for more oversight and the desire of the PMs for less work. Many felt that those in OSD with responsibility for overseeing the technical and other risks of the program should be accountable for the program. In other words, if the program has cost and schedule growth, the responsibility for this growth would be shared by both the program office and the OSD stakeholder. To mitigate variability in interpreting requirements, some suggested establishing a minimum threshold for DoDI 5000.02 document requirements. Obtaining information above the threshold would require the MDA approval.

Summary

Both OSD and the Navy program management communities felt that ships are different from other MDAPs. Reasons they cited for this included that ships take longer to design and build, have concurrent design and build processes, have an influential political and industrial base, are more complex, are produced at relatively low rates and quantities, and have high unit costs and a unique funding construct. We attempt to validate and identify the implications of these and other perceived differences between ships and other MDAPs in the next chapter.

Nevertheless, interviewees differed in the primary issues or challenges they identified for the 5000 process. OSD interviewees were most concerned with the inability to prototype and test a unit before having multiple ships on contract. They were concerned with the continuous revisiting of the same issues, resulting from ambiguity and interpretation of the 5000's "start of production," "LRIP," "FRP," and meaning of MS B and MS C for ships. The Navy program management community was primarily concerned with the timing and content of the numerous documents required in the 5000 process. Many felt that the documents did not add value to the DoD or the PM and required detail beyond the maturity of the information available at the time the document is required. They also felt that the ambiguity surrounding the start of production, LRIP, and FRP, and the meaning of MS B and MS C, were problematic. How to accommodate new prototyping rules and technology readiness requirements by MS B was also identified as an important issue. Both communities felt that the testing requirements and approach needed revision.

Although the issues differed somewhat between the OSD and PM communities, the solutions proposed by each were similar. Most felt that improving the ability to tailor, rethinking the meaning of currently ambiguous definitions for ships (LRIP, FRP, MS B, and MS C), rethinking the best way to test and evaluate ships, and generating an annex to the 5000 to clarify the currently ambiguous or sticking points for ships would all be helpful.

CHAPTER FOUR

Program Comparisons

The stakeholder interviews summarized in the previous chapter identified many characteristics of ship programs that are perceived to differ from those for other complex MDAPs. In this chapter, we examine whether program data and histories validate these differences, what their process implications might be, and how ship programs might differ from one another.

We examine both ship and nonship programs to identify what may drive differences in program structure, workflow, and execution. We identify and analyze metrics or indicators that reflect technical/engineering or programmatic differences. We begin by discussing data sources and program samples. We then show a time line comparison of the major events for four ship programs: the SSN-688, DDG-51, SSN-774, and DDG-1000. This comparison shows how the acquisition process differs across ship classes and over successive generations. Next, we compare the characteristics of a set of ship and nonship programs to illustrate the differences in the timing and duration of key acquisition events. We also discuss potential implications of these differences for managing and tailoring the acquisition process. Last, we summarize the differences we observe.

Approach and Data

RAND collected data on a number of ship and nonship programs to identify how ships may differ from other major weapon-system programs. We collected data to compare acquisition time lines and major activities. We also collected data to evaluate specific issues, such as the small number of units procured and challenges associated with test and evaluation activities, which our interviewees identified as affecting shipbuilding programs.

The U.S. Navy procures many types of ships, including aircraft carriers, submarines, auxiliary ships, surface combatants, and amphibious ships. Other MDAPs also span a wide range of system types, including aircraft, missiles, tanks, information systems, and satellites, as well as other services. We selected our sample programs to cover most but not all of these categories. However because of time and data limitations, our actual comparison sample was more limited in scope. Table 4.1 summarizes the ship and nonship programs we examined. Appendix B lists all programs we considered and the data that were available for them; programs not selected typically required more time or data than available for analysis.

The programs in our sample are at various points in the acquisition life cycle and may have had significant milestones achieved under previous versions of the DoDI 5000-series regulations. For example, CVN-21 has no construction experience to date, while DDG-1000 (the lead ship) is approximately 50 percent complete. Other programs, for example, SSN-688 and DDG-51, were well into their production and deployment phases.

The information for describing and comparing ship and nonship program structures and acquisition strategies came from multiple sources, including SARs and Defense Acquisition Executive Summary reports for each program, available official program documentation obtained in the course of prior RAND work, and approved acquisition strategies.

Table 4.1
Programs Included in Comparative Sample

	Program	Program Description	Service	Date of MS B/II
Ship	SSN-688	Submarine	Navy	No MS B/II[a]
Ship	DDG-51	Surface combatant	Navy	Dec 1983
Ship	SSN-774	Submarine	Navy	Jun 1995
Ship	LPD-17	Amphibious	Navy	Jun 1996
Ship	T-AKE	Auxiliary	Navy	Oct 2001[b]
Ship	CVN-21	Aircraft carrier	Navy	Apr 2004
Ship	DDG-1000	Surface combatant	Navy	Nov 2005
Nonship	AMRAAM	Missile	Air Force	Sep 1982
Nonship	C-17	Aircraft (heavy lift)	Air Force	Feb 1985
Nonship	SBIRS (High)	Satellite	Air Force	Oct 1996
Nonship	SM-6	Missile	Navy	Jun 2004
Nonship	F-22	Tactical aircraft	Air Force	Jun 1991

[a] SSN-668 did not have an MS II nor did it have an equivalent. The program went from MS I to MS III in only three years.

[b] The T-AKE program also skipped MS B; however, the equivalent of a MS B is the lead ship detail design and construction contract from October 2001.

Ship Program Structures

One important consideration in this research was the extent to which process tailoring took place to enable a better fit between ship programs and the acquisition process. Process tailoring is allowed per DoDI 5000.02 with approval of the MDA for the program.[1] Tailoring includes the type and timing of technical reviews, milestones, and other oversight activities; the timing and scope of contract awards; and test activities. The combination of these categories of program events and activities defines the program phases and the transition from one phase to the next. Tailoring may also include the use of competition, the use of prototypes or other preproduction configuration models used for demonstration and test, the type of contracts awarded, and milestone entry and exit criteria. All these program descriptors are usually documented in a program's Acquisition Strategy, which is required at MS B and C and signed by the PM, PEO, SAE, and DAE.

Figures 4.1 through 4.4 present the program structure of four ship programs in two classes—destroyers and attack submarines. Each figure shows the program time line, including the dates of milestones, contract awards, ship deliveries, and test events. Although these figures are complex, they illuminate the sequence of key events and the timing between them. One important observation is that the relationships among the various events and activities shown tend to be unique to each program. None of the ship programs we examined had exactly the same program structure, reflecting the unique attributes of each program. Thus, ship programs are not necessarily a homogeneous set of program structures; rather, they are significantly different from each other, reflecting differences in technical characteristics and maturity, contracting approaches, and regulatory regimes.

A simple comparison of these four programs is instructive in a number of dimensions. Changes from one generation to the next within a ship class highlight changes not only in regulatory regime over time but also in the design and build process. Looking across the two ship classes, the relative uniqueness of each program structure is readily apparent, but several aspects of that structure appear to be common across all four programs (and are also common to the other ship program structures described in Appendix B).

DoD was just beginning to adopt the set of policies and procedures that constitute the acquisition process as we know it today when the Los Angeles Class attack submarine program (SSN 688) began. Figure 4.1 shows the compressed early phases and single production decision very early in the program that make this program distinct. A multiship follow-on contract was awarded only one year after the lead ship design contract. Electric Boat and Newport News, the two shipyards capable of designing and

---

[1]   The MDA for ACAT 1D programs and designated special interest programs is the USD (AT&L) in the role of Defense Acquisition Executive (DAE). For ACAT 1C programs (and most other programs), the MDA is the Service Acquisition Executive (SAE), usually an assistant secretary.

Figure 4.1
Los Angeles Class SSN-688 Submarine Program Structure

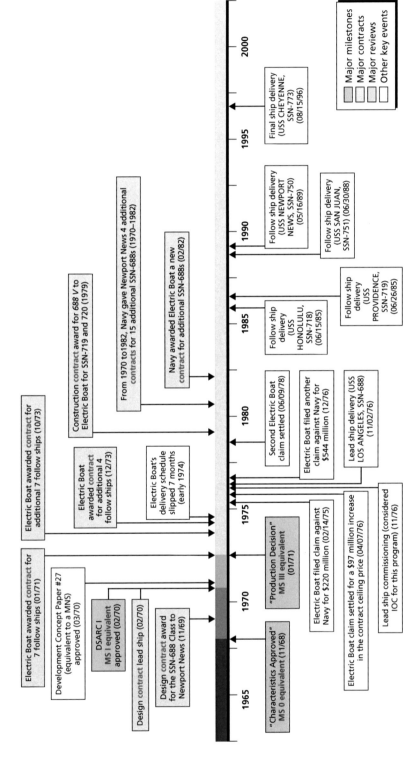

NOTE: Blue shading on time line indicates length of MS 0, MS I, and MS III.

RAND MG991-4.1

constructing submarines, received a series of multiyear, multiship construction contracts. Eighteen ships were on contract before the delivery of the first ship 81 months after MS I.

The later Virginia Class attack submarine program (SSN-774) also had fairly compressed early program phases and multiyear, multiship contract awards, as Figure 4.2 depicts, but other aspects of the program were very different. The Virginia Class had a longer LRIP phase and additional production decisions not present in the Los Angeles Class program. The Virginia Class used an integrated product and process development approach as the foundation for its design/build strategy, enabled by computer design tools not available 30 years earlier. Milestone II included LRIP approval, a PDR was conducted after MS II, and 18 ships were on contract before the start of IOT&E over three years after first delivery. Eleven ships were on contract before the first delivery 122 months after MS I.

The Arleigh Burke Class guided missile destroyer program (DDG-51) began chronologically in between the two submarine programs. The initial program phases, as defined by the major milestones, were again somewhat compressed. Program phases were relatively short, with only six years between MS 0 and MS III. The lead ship was approved at MS II, in the middle of contract design, with detail design and lead ship construction contract award occurring 16 months later. Seventeen ships (27 percent of the eventual total quantity) were on contract before operational evaluation (OPEVAL) completion. The DDG-51 program was also broken into three distinct configurations, called flights, each incorporating different or updated capabilities. For a modern ship, DDG-51 had two characteristics that are uncommon today: (1) a production rate that varied above one per year for a number of years, and (2) a relatively large total quantity of 62 ships. The time from MS I to lead ship delivery of 118 months, and the time from MS I until initial operational capability of 140 months, are both comparable to the time the Virginia Class needed for these metrics (122 months and 152 months, respectively).

The Zumwalt Class destroyer program (DDG-1000) is one of the Navy's recent surface combatants. As Figure 4.4 illustrates, the program had what might be considered a more traditional structure, with longer phases, milestones more spread out, and technical reviews common to most MDAPs. The DDG-1000 program evolved over a long period of time, with relatively long phases and 20 years between MS 0 and the planned MS C. Detail design was approved at MS B and LRIP quantities authorized. Nevertheless, the lead ship construction contract was split from the detail design award, requiring an additional DAE program review before contract award for the lead ships. Similar to SSN-774, industry base considerations played an important role in the program structure and execution. In this case, the Acquisition Strategy approved at MS B called for "dual" lead ships to alleviate variation in the workload at the two shipyards capable of designing and building this type of ship. Under the program of

Figure 4.2
Virginia Class SSN-774 Submarine Program Structure

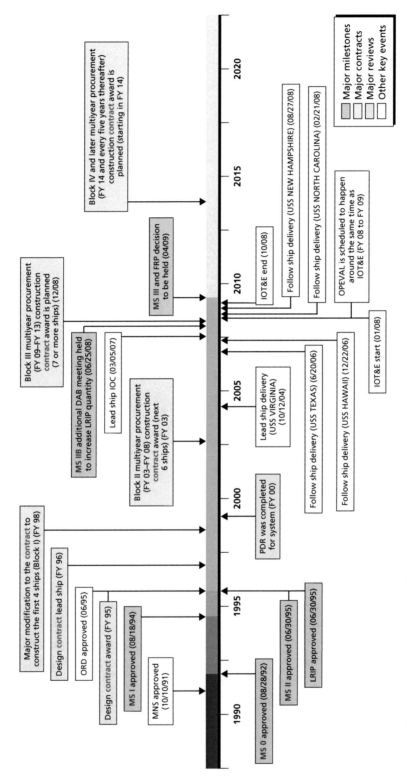

NOTES: Blue shading on time line indicates length of MS 0, MS I, MS II, and MS III. Date for critical design review for submarine system is not available.

RAND MG991-4.2

Figure 4.3
USS ARLEIGH BURKE DDG-51 Destroyer Program Structure

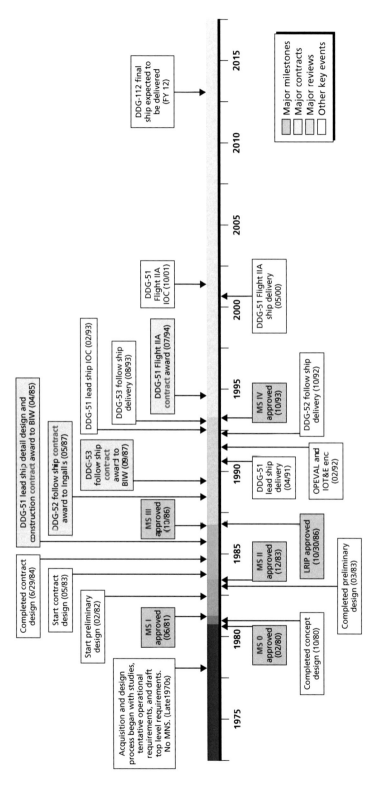

NOTES: Blue shading on time line indicates length of MS 0, MS I, MS II, MS III, and MS IV. Flight I Hulls 51–71 were delivered between April 1991 and August 1997. Flight II Hulls 72–78 were delivered between August 1997 and January 1999. Flight IIA Hulls 79–112 were or will be delivered between May 2000 and 2011 or 2012. Dates for preliminary design review and critical design review for ship system are not available.

RAND *MG991-4.3*

Figure 4.4
USS ZUMWALT DDG-1000 Destroyer Program Structure

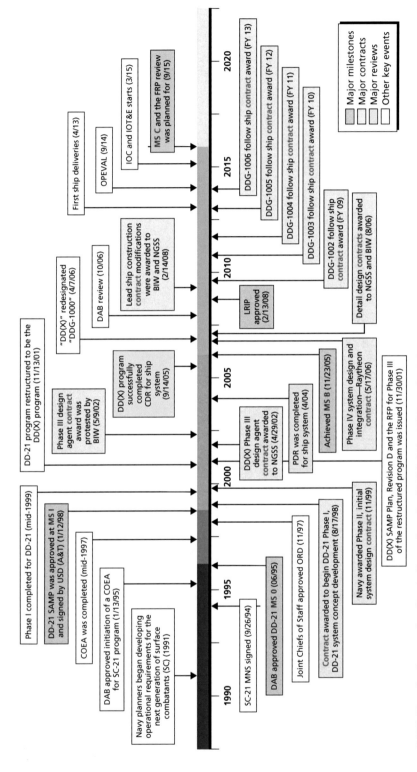

NOTE: Blue shading on time line indicates length of MS 0, MS I, MS B, and MS C.

RAND MG991-4.4

record during 2008, all seven planned ships were planned to be on contract before OPEVAL started.[2]

These program examples illustrate some of the unique attributes of ship programs raised during discussions with Navy and OSD stakeholders, including

- compressed early program phases
- program phases defined more by contract awards than by major milestones
- contract awards defined by the sequence of activities in the ship design/build process
- relatively small total quantities
- low annual production rates, sometimes only one per year
- significant portion of total quantity on contract before testing of the lead ship is complete (concurrency)
- a significant role for the industrial base in terms of influencing program structure and contracting strategy.

This last item is an important attribute of many ship programs. The very large workforce required to construct large, complex ships, and the need to keep those workforce levels as stable as possible, means that industry base considerations play an explicit role when formulating a ship program's Acquisition Strategy.

Perhaps most important, there appears to be a mismatch between major milestones and such key program events as contract awards and testing. That is, ship program design/build events do not appear to align well with the intent, timing, scope, and content of some milestone reviews. Often, several contract awards denote different design stages (e.g., system design, functional design, contract design) before MS B (or II) rather than technology development and demonstration. MS B tends to not only approve continued design activity (detail design) but also initial production (e.g., the lead ship). Under DoDI 5000.02, MS B is intended as the start of product development—integrating technologies and maturing concepts into a form intended for deployment to the warfighter, whereas ship programs tend to treat this milestone as a continuation of design activities. With some notable exceptions, system and subsystem demonstration through testing must wait for delivery and test of the lead ship, as opposed to the construction of developmental systems intended only for test.[3]

Although ship programs do have the elements listed above in common as a class of weapon systems, they can also be very different from each other. The exact sequence, timing, and scope of contract awards can be different as a result of ship type, size, design and technology maturity, roles and responsibilities of government and industry,

---

[2] The Secretary of Defense truncated the program at three ships during the FY 2010 budget cycle, opting instead to restart DDG 51 production.

[3] The DDG 1000 program (then known as DD(X)) managed, and benefited from, a series of major subsystem and critical technology tests using Engineering Development Models (EDMs) before MS B.

preferred design tools, and industrial base characteristics. The presence and timing of major decision milestones, technical reviews, and test events can differ among ship types for similar reasons.

Ship and Nonship Program Comparisons

DoD acquires many different types of weapon systems. The largest and often the most technically complex are ACAT I programs, which, in addition to ships, include space systems, aircraft, missiles, and ground vehicles. Despite obvious differences in physical characteristics, missions, and expected operational environments, all ACAT I programs fall under DoD 5000-series regulations. As mentioned, the regulations allow tailoring of the process to meet the needs of a specific program. The notion that all large complex systems can benefit from similar design, development, production, and oversight processes, as long as there is some flexibility to tailor to unique circumstances, has underlain acquisition policy and regulations for almost four decades. The benefits of having a single process, or at least very few processes that are similar in intent and language, was recently explored and validated by an internal OUSD (AT&L) initiative.[4]

Figure 4.5 illustrates one issue in using generic models of the acquisition process as defined in DoDI 5000.02 and a generic ship design/build model. The figure shows a highly simplified view of the design process for ships. The top portion of the figure represents the DoD 5000 process, and the bottom portion of the figure identifies where the traditional ship development phases align to the 5000 process. The majority of the early design work occurs during the technology development phase. Detail design and construction activities occur during the EMD phase. For modern ships, these design phases overlap to such an extent such that it is difficult to define when, for example, preliminary design ends and contract design begins.[5] Note also that ship design phases and associated contract awards and technical reviews do not always correspond with the traditional model. Nevertheless, MS A and MS B still occur at transition points between phases. It is really MS C that has no corresponding equivalent in the ship design/build model.

Table 4.2 shows the basic technical characteristics of three contemporary MDAPs. From this perspective, it is abundantly clear that a submarine (SSBN) is very different from a fixed-wing fighter aircraft (F-16) or a tank (M-1). The ship is much larger

---

[4]   Department of Defense, Office of the Under Secretary of Defense, Acquisition, Technology, and Logistics, 2008a (draft version 8), and 2009.

[5]   In Chapter Five, we revisit the ship design process in Figure 5.1.

Figure 4.5
Notional Ship Program Phases Within the Acquisition Process

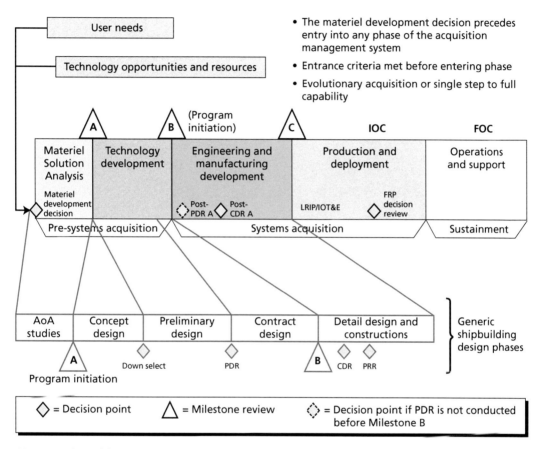

SOURCE: Adapted from Fireman, 2007.
RAND MG991-4.5

in both weight and size, has ten times more subsystems and components,[6] has a much larger crew, has longer unit production time and associated labor-hours to produce, has many more suppliers, and has a lower production rate. Although these kinds of differences are obvious in any comparison of ship and nonship programs, the question is whether those differences warrant the use of a different set of acquisition policies and practices.

Table 4.3 compares a few additional ship and nonship programs across a set of metrics on each program's Acquisition Strategy. Program initiation, here defined as the

---

[6] Note that the numbers of "parts" and "suppliers" are exceedingly difficult to count on an equivalent basis across systems. These numbers should be taken as notional only.

Table 4.2
Comparison of Technical Characteristics for Three Programs

	Ohio Class SSBN	F-16	M-1
Weight (tons)	18,750	10	65
Length (ft)	560	49	26
No. of subsystems	265	32	26
No. of components	25,000	28	212
Patrol/sortie duration	3 months	1.4 hrs	1 day
Operational life	30 yrs (now 44)	8,000 hrs	20–30 yrs
Crew size	150	1	4
Unit production time (months)	72	32	7.5
No. of part numbers	350,000	175,000	14,065
No. of suppliers	4,500	850	600
No. of man-hours/unit	12 million	57.5 thousand	5.5 thousand
Annual production rate	1	200+	600

SOURCE: Naval Sea Systems Command, 1992.

first major milestone in which a system concept and coherent program was presented to senior decisionmakers, suggests that ship programs can start earlier than nonship programs.[7] This does not mean that ship programs have a more mature design and technology set earlier in the acquisition cycle, only that program planning might be further along. There is no milestone equivalent to MS 0 in today's acquisition process; however, the materiel development decision is similar in intent.

Unit costs tend to be much higher for ships than for most other ACAT I programs, except satellites. There are more ambiguous differences between ships and other program in the proportion of total program funding that is RDT&E. Some interviewees said that ships tend to have relatively smaller proportions of funding in RDT&E, in part because shipboard systems are often developed outside the ship program, and in part because the first ship uses procurement funds (SCN) rather than the RDT&E funds. The lead ship in a ship program, even though it uses SCN funding, also includes substantial design and technology development activities usually funded with RDT&E in nonship programs.

---

[7] DoDI 5000.02 defines formal program initiation at MS B. The regulation allows program initiation for ships at MS A, but only the LCS program has launched initiation then. The underlying issue here is how mature the concept, technology, and design should be at program initiation. A formal program is expected to be more mature and so able to offer more credible and detailed program information, including risk assessments and cost and schedule estimates.

Table 4.3
Comparison of Select Program Descriptors

Difference	Metric	Ships					Other MDAPs				
		DDG-51	LPD-17	SSN-774	DDG-1000	CVN-21	AMRAAM	C-17	SBIRS (High)	SM-6	F-22
Program initiation[a]	First milestone	MS 0 Feb 80	MS 0 Nov 90	MS 0 Aug 92	MS 0 Jun 95	MS 0 Mar 96	MS I Nov 78	MS II Feb 85	MS II Oct 96	MS B Jun 04	MS I Oct 86
Unit cost	PAUC ($M BY 2006)	$1,084	$1,352	$2,536	$3,659	$9,307	$1.18	$360	$2,824	$4.07	$364.1
R&D funding[e]	% RDT&E funding	6	1	8	35	14	18	16	77	18	50.5
Quantity	Planned quantity	62	9	30	7	3	13,953	190	4	1,200	184
Production rate	LRIP quantity	9	No LRIP	18	7	3	4,159	40	No LRIP	120	25
	Annual rate (min/max)	1/5	0/2	1/2	1/1	0/1	97/1,165	2/22	0/1	20/150	2/24
Phase time	MS I/A to MS II/B (months)	30	42	11	96	47	47	32[b]	No MS A[c]	No MS A[d]	57
	MS B to 1st delivery (months)	91	110	113	89	141	74	69	95	66	146

SOURCE: Data are from SARs and program Acquisition Strategies. Data for some programs represent the current estimates rather than actual values.

[a] Program initiation is officially MS II or MS B, according to DoDI 5000.2. Here, we mean the first milestone in which a coherent program is presented to decisionmakers.

[b] In the absence of an official MS I for the C-17, we used the date when the research and development contract was awarded (July 1982).

[c] There is no MS A or an equivalent available for SBIRS (High).

[d] There is no MS A or an equivalent available for SM-6.

[e] Percentages are estimated from total program cost and RDT&E cost as reported in the December 2007 *Selected Acquisition Report* for that particular program.

One ship program, DDG-1000, had a very high proportion of RDT&E funds, whereas another, CVN-21, is comparable to AMRAAM, C-17, and SM-6. The DDG-1000 and CVN-21 programs both included development and test of major subsystems and associated technology; in the past, such efforts were often managed separately outside the ship program office. For example, the Aegis system, which provides air combat and air defense capabilities to both CG-47 and DDG-51 ship classes, was managed as a separate program within NAVSEA. The other ship programs shown in Table 4.3 had smaller RDT&E proportions, and the satellite program had a very high proportion of RDT&E funds. The results for DDG-1000 and SBIRS (High) may be driven by the low quantity and, hence, relatively low procurement funding, in each program, although both systems also sought state-of-the-art advances in their concepts and technologies, requiring a substantial development effort.

In general, ship programs have lower LRIP quantities and lower planned total quantities than other weapon systems, except satellites. Annual production rates for ships vary from less than one (carriers) to five (DDG-51), but for recent ship programs, production rates are more often one per year. Nonship programs have a wider range of annual production rates, from none (satellites) to hundreds or even thousands (missiles). The relatively longer time and higher unit cost it takes to produce a complex ship (or satellite) contributes to their lower production rates.

Table 4.3 also shows the duration of two program phases: technology development and engineering and manufacturing development. Technology development is the time between MS I or A and MS II or B. As mentioned above, ship programs tend to have formal early milestones (MS 0, MS I, or MS A); among the small sample of nonship programs we examined here, most did not have a MS I/A. Yet, it is noteworthy that the technology development phase was highly variable among ship programs, and the one nonship program for which we had data (AMRAAM) fell into the middle of this range.

The results for the EMD phase—measured as the time between MS II or B and first delivery—suggest a significant difference between ship and nonship programs. The EMD phase for ship programs is fairly long, ranging, for the programs in our sample, from 91 months for surface combatants to 141 months for the carrier. Nonship EMD phases range from 66 to 95 months, with the longest duration for the satellite, suggesting some important program structure similarities between satellites and ships. The activities taking place during this phase are also notable. For ship programs, EMD is detail design and lead ship construction. For nonship programs, EMD is system (or product) design, production planning, and production of developmental test articles.

Two important policy issues frequently raised in the stakeholder discussions were the timing of test and evaluation activities and the meaning (or scope) of major decision milestones. Table 4.4 presents some schedule-based metrics that highlight these issues and show some differences between ship and other programs. For ships, the time between the initiation of LRIP (often at MS B with approval of the lead ship, or the

Table 4.4
Specific Issue Comparison: T&E Timing and Meaning of Key Milestones

Difference	Metric	Ships					Other MDAPs				
		DDG-51	LPD-17	SSN-774	DDG-1000	CVN-21	AMRAAM	C-17	SBIRS (High)	SM-6	F-22
T&E timing	LRIP to IOT&E end (months)	64	No LRIP	161	86	174[a]	19	76	No LRIP	75[a]	40.6
	LRIP to 1st delivery (months)	55	No LRIP	113	62	141	16	18	No LRIP	63	22
	1st follow-on unit award to IOT&E end (months)	58	120	135	79	84[a]	20	NA[b]	No data	73[b]	40
Scope of milestone	% planned quantity on contract by MS C	2	100	33	100	100	17	1	No MS C[c]	0	1.6
	LRIP quantity as a % of total	15	No LRIP	60	100	100	30	21	No LRIP	10	14

[a] Used OPEVAL in the absence of IOT&E data.
[b] IOT&E ended before the follow-on was awarded.
[c] There is no MS C or an equivalent available for SBIRS (High).

award of the lead ship construction contract) and the completion of IOT&E tends to be quite long. This reflects both the long time it takes to construct a complex ship and the time it takes to actually test the lead ship. IOT&E requires the use of a production representative unit; for ships, that unit is the lead ship. LRIP approval and MS B do not correspond in most nonship programs. LRIP is initiated at MS C (per DoD 5000.02) or at a point late in the EMD phase. As a result, the time between the start of LRIP and the completion of IOT&E tends to be shorter. The time between LRIP start to first delivery shows the same pattern. The time between the first follow ship award and completion of IOT&E is also lengthy for ships, indicating that follow-on contract award occurs significantly earlier than delivery and test of the first ship. We saw this pattern in the ship program structures discussed in the previous subsection.

The meaning or scope of the major milestones is more ambiguous for ships than for other programs. As mentioned above, ships have the option of formal program initiation at MS A, but DoDI 5000.02 does not specify this option for other programs. For ships, MS B can approve the award of the detail design and lead ship construction contract, which might be considered the start of LRIP. For nonship programs, MS C is usually considered the decision point for LRIP. Table 4.4 illustrates this difference for ship programs. A significant percentage of the total planned number of production units for ship programs is already on contract by MS C—in some cases, the entire planned production quantity. This is generally not true for nonship programs. The result is that MS C for ships is highly ambiguous. At a minimum, it will not mean the same thing, or cover the same set of issues, as MS C for nonship programs.

Some differences between ship and nonship programs do not have clear metrics but nevertheless appear to be real. These include

- The timing of information requirements supporting milestone decisions and the technical activities that generate that information. Because of the long ship design/build time lines, the technical reviews or tests that would give decision-makers critical information do not occur until after the planned milestone. One manifestation of this is that milestone documentation requirements can sometimes include a level of detail not available in ship programs until later.
- The role of the industrial base in ship programs is more significant than in nonship programs. This influence manifests in contracting strategies, the timing of contracts, the explicit consideration of shipyard workload as a factor affecting Acquisition Strategy decisions, and the lack of competition (or increased collaboration) between shipyards. Ship program review decision briefings often include proprietary workload charts for the participating shipyards, showing the implications of contract award or construction approval timing. Such explicit consider-

ation of labor workload issues was very unusual in nonship programs until the early 1990s.[8]

- Long ship design/build time lines presenting a problem of technological obsolescence. Although this problem is also present in some nonship programs with long developmental phases, it may be felt to a higher degree in ship programs and across more subsystems on the ship.

- Inappropriateness of live fire test activities risking damage to the lead ship, a deployable asset. Nevertheless, this issue is increasingly present in many MDAPs as the cost of individual articles, even dedicated test articles, argues against destruction.

Ships and Satellites

Satellites share some attributes that make ship programs a poor fit with some aspects of the traditional acquisition process model. These shared characteristics include complex system with a long design/build time frame, first unit is deployed and operational, testing cannot be completed until after delivery of the first unit, potentially small total quantities, and low annual production rates.

Although ships and satellites share these unique attributes, they are subject to different acquisition processes. The acquisition process for satellites is described in the *National Security Space Acquisition Policy: Interim Guidance for DoD Space System Acquisition Process* (2009). The guidance for satellites describes two acquisition processes, based on the number of units to be produced. The Small Quantity System Model is typically used for quantities of ten or less, and the Large Quantity Production Focused Model is typically used for systems procured in quantities of 50 or more. Each process consists of the same four acquisition phases: Phase A, Concept Development; Phase B, Preliminary Design; Phase C, Complete Design; and Phase D, Build and Operations. The main difference between these two processes is that, for programs with larger quantities, there is an LRIP decision and full rate initial production decision instead of the build approval, follow-on buy approval, and upgrade decisions that occur for smaller quantities. For large quantity programs, there is also, potentially, a system demonstration subphase where prototypes may be demonstrated.

The placement of milestones to create a program structure may differ between the two models. Figure 4.6 and Figure 4.7 depict the acquisition process guidance outlined

---

[8] Industry consolidation, procurement budget decreases within some commodity types, and a decrease in the frequency of new program starts have highlighted workforce issues in aircraft, helicopter, and heavy armored vehicles. See Drezner et al., 1992, and Birkler et al., 2003, for further discussion and examples.

Figure 4.6
Interim National Security Space Acquisition Phases: Small Quantity Production

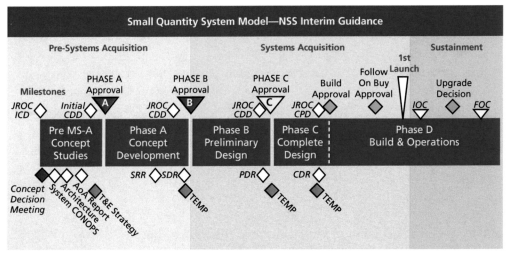

SOURCE: *National Security Space Acquisition Policy: Interim Guidance for DoD Space System Acquisition Process,* Figure AP1-1, 2009, p. 7.

RAND *MG991-4.6*

Figure 4.7
Interim National Security Space Acquisition Phases: Large Quantity Production

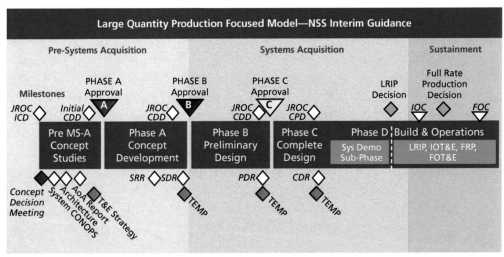

SOURCE: *National Security Space Acquisition Policy: Interim Guidance for DoD Space System Acquisition Process,* Figure AP1-2, 2009, p. 8.

RAND *MG991-4.7*

for the Small Quantity System Model and Large Quantity Production Focused Model, respectively.

There are some notable differences between these processes and the traditional acquisition process outlined in DoDI 5000.02. The first and obvious difference is that satellites have two processes to choose from, depending on the expected production quantity. The large quantity acquisition model more closely aligns with the process outlined in 5000.02. The small quantity process has many more unique attributes and may more closely align with historical ship acquisition program structures and processes. The majority of the acquisition phases for satellites appear to be focused more on maturing the design of the system than on the development and test of systems than in the 5000.02 process. For satellites, many requirements depend on the characteristics of the item being procured. For example, testing critical technologies in a relevant environment is required only when feasible. LRIP and FRP decisions are required only if there will be many units produced. In both models, the Phase C decision point roughly aligns with MS B under DoDI 5000.02. Similarly, PDR occurs before the MS B (or Phase C) decision, and CDR occurs before the ship (or satellite) build approval.

Satellite programs also have unique characteristics, such as a small, highly skilled workforce, very densely packed operational units, and the unique operational environment of space. The implication is that a program structure tailored specifically for a complex ship will not be appropriate for a satellite, and vice versa. This leads us back to the notion that acquisition processes need to be tailored to the unique attributes of the system of interest—a fundamental underpinning of DoD 5000-series regulations governing the acquisition process.

Summary

The differences in program activities between ship and other programs (or the traditional acquisition framework) have implications for the fit between program structure and formal policy and process, as well as the ability to tailor that process. For instance, MS C is not just a production milestone. It also signals the completion of development. In practice, this translated to EMD completion under the revised, 2008, DoDI 5000.02 (or SDD complete under the 2003 DoDI 5000.2) as one entrance criterion for MS C. EMD (formally SDD) is not officially complete until IOT&E is complete, which is also necessary for the Beyond LRIP report required for an FRP decision. Many ship programs will have a significant proportion of the total planned quantity on contract before delivery of the lead ship and completion of IOT&E. Because annual production rates do not change for many complex ships, the normal distinction between low-rate and full-rate production is not relevant. Nevertheless, dropping MS C for ships would risk losing its other nonproduction-related attributes.

However, ship acquisition programs do not appear to be so unique that all the principles and processes in DoDI 5000.02 do not apply. Historical ship program structures demonstrate that it is possible to align portions of the ship program, particularly the early phases, with the more traditional acquisition process applied to MDAPs. Careful tailoring of the program structure to meet a ship program's needs in terms of contracting strategy, technology development, and early design activities is feasible.

What is missing from DoDI 5000.02 and its associated guidance (e.g., the *Defense Acquisition Guidebook*) is a framework or guidance that helps acquisition process stakeholders, from the PM to OUSD (AT&L), identify the attributes of a program that require tailoring and gain some notion of what appropriate tailoring would look like (i.e., acceptable boundaries).

CHAPTER FIVE

Ship Acquisition Alternatives

As earlier chapters showed, many of the differences attributed to ship programs relate to decision points and oversight processes and the availability of information associated with technical activities, including testing. In this chapter, we explore how these key decision points (e.g., milestones) may be tailored for ship programs. There are many possible places for oversight activities and key decision points in the DoDI 5000 acquisition framework. Several variables influence the timing and scope of these activities. For example, the design/build process,[1] statutory and regulatory constraints, and technical aspects of the program can all dictate what oversight activities will occur, and when. The nature and timing of these decision points depend, at least in part, on the type of system being procured.

We begin by describing the current design/build process and oversight activities for ships. We then explore the variables that affect the timing and scope of oversight activities and outline a set of acquisition alternatives for ships. The first set of alternatives assumes that the process is unconstrained—a thought exercise to understand the range of possible alternatives. For the second set of alternatives introduced, we explore how constraints affect the timing and scope of oversight activities.

The Current Design/Build and Oversight Process

The design/build process for ships has evolved over time to become an increasingly seamless process, with notable overlap between the design and construction phases. This process has implications for oversight and the application of the DoD 5000 process. Figure 5.1 shows a simplified version of the major activities in the ship acquisition process, to include the design/build process.[2] Figure 5.1 also shows the points at which the primary oversight activities and milestones typically occur.

---

[1] By *design/build process*, we mean the technical, engineering, and construction activities required to take a ship from concept to an operating system.

[2] Note that this figure is very different from Figure 4.5. Rather than showing the ship design/build process in context of the DoDI 5000.02 process, we are highlighting the complexity and activities associated with building and designing ships. We have noted key milestones and decision points as they notionally occur.

Figure 5.1
Ship Design/Build and Oversight Process

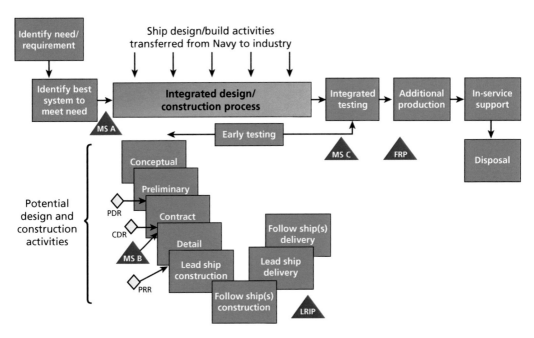

As we have discussed, the current ship acquisition process typically begins with the identification of a requirement and analyses to identify a system solution. This is the case for both ship and nonship programs. Under the revised, 2008 DoDI 5000.02, a materiel development decision (not shown on the chart) would approve continuation of exploration for a system solution and direct entry into the acquisition process at a point commensurate with concept, technology, and system design maturity. At Milestone A, the technology development phase begins and system design and technology development work begins and progresses until both are mature enough to initiate the engineering and manufacturing development phase at Milestone B. For ships, Milestone B typically occurs before the detail design and construction contract award but after the preliminary and critical design reviews. Low-rate initial production quantities are typically authorized at MS B, and contracts for follow ships are let sometime after lead ship construction has commenced. Subsystem testing can begin as early as the design phase, and system-level testing is completed some years after the delivery of the first ship. Milestone C occurs when testing is near complete. If additional follow ships are required at that time, a full-rate production decision can be made.

Figure 5.1 shows a typical placement of DoDI 5000.02 milestones and OSD reviews for the ship acquisition process. Although there are currently a number of challenges with determining when these particular activities should occur for ship programs,[3] it is possible to establish a meaningful set of oversight activities and OSD decision points for ships. Figure 5.2 identifies a generic set of oversight activities and key decision points and where they could occur in the ship design/build process.

This set of oversight activities and decision points is not intended to be comprehensive but to represent the key activities and decision points in the ship acquisition process. For nearly every major activity in the ship acquisition process, there is (or can be) a corresponding OSD evaluation. The major decision points revolve around contract awards and production decisions.

We did not assign these decision points or activities to specific milestones, because there are many options for placement. A number of variables, which we describe below, influence the placement of OSD decision points and oversight activities. Figure 5.3 shows the possible assignment of some of these activities to milestones. The evaluation of design maturity and contract award could occur at any milestone. Evaluation of construction readiness and contract award could occur at MS B, MS C, or other decision points, to include interim progress reviews (IPR) before MS B. The evaluation of ship ability to meet requirements can occur at MS B, MS C, or other decision points. Follow-on construction decisions could occur at MS C or other decision points.

Figure 5.2
OSD Oversight Activities and Decision Points in the Ship Design/Build Process, Developed by RAND

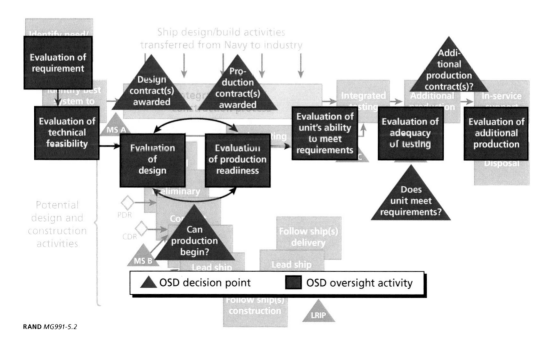

RAND *MG991-5.2*

---

[3] See Chapters Two and Three.

Figure 5.3
Potential Linking of Activities to Milestones

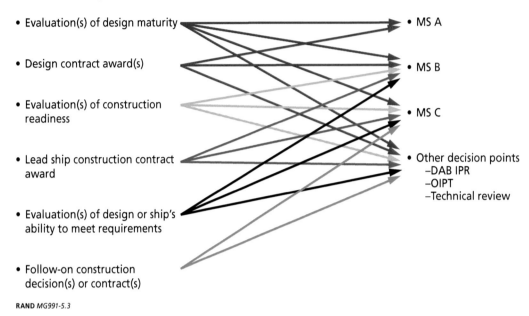

In an unconstrained world, the Navy can control programmatic decision variables affecting the linking of these activities. For example, the Navy can determine the development phase length, the level of overlap between phases, and the duration of the gap between lead and follow ship. The Navy can also determine the level of technical risk assumed (e.g., by how much design and construction of the lead ship is overlapped, or design maturity), the level of competition to be employed, and the level of prototyping or engineering development model testing. It can also influence the choice of design and manufacturing tools. All this will affect the timing and scope of oversight activities and decisions.

Unconstrained Ship Design/Build and Oversight Process Alternatives

Assuming that the Navy has the ability to control the variables above, it has numerous potential ship design/build and oversight processes. Ship acquisition programs tend to have a unique design/build process reflecting the technical characteristics and engineering challenges of each ship. The following examples serve to illuminate the range of possibilities.

The ship design/build process is largely defined by the level of overlap between the development, design, and construction phases. Figure 5.4 illustrates three possible ship design/build constructs, including no overlap of the phases; overlap of the design and construction phases only; or overlap of the technology development, design, and construction phases.

Figure 5.4
Possible Design/Build Processes

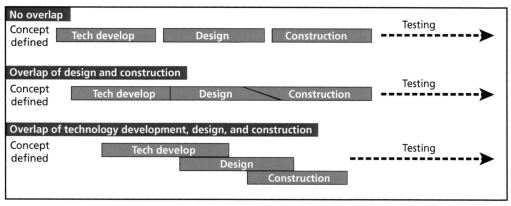

RAND *MG991-5.4*

There are pros and cons to each of these constructs. The "no overlap" design-build construct may reduce certain technical risks and production rework and redesign resulting from immature technologies or an immature design. In this alternative, the technology must be fully developed before the start of design, and the design must be complete before the start of construction. However, the "no overlap" approach takes relatively longer and increases the risks associated with technical obsolescence and with changes to the threats driving requirements. It also might impose more risk on the industrial base because it makes it more difficult to maintain continuity of work. As the level of overlap between phases increases, technical and production risks increase, whereas technology obsolescence, requirements change, and industry base risks decrease. A compressed schedule can allow for earlier fielding of technology, if the technology is already mature.

In each of these design/build constructs, key activities, such as major milestones, could occur early or late in the process or could be allowed to vary. Depending on when these activities occur, their scope or meaning can differ. Table 5.1 defines the oversight activities for each of the "early," "late," and "variable" milestone constructs and introduces IPRs for important decision points otherwise associated with a milestone. In each construct, Milestone A occurs at the start of the technology development phase. In the "early" construct, Milestone B occurs at the start of the detail design. Lead ship construction is authorized at Milestone C, and an IPR occurs on the completion of testing of the first ship to determine follow-on construction units. In the "late" construct, the start of detail design (and subsequent contract) occurs at an IPR after Milestone A but before Milestone B. At Milestone B, construction of the lead ship is authorized. Milestone C occurs when lead ship construction is authorized. An IPR is

Table 5.1
Possible Timing and Scope of Milestones, IPR

Early	MS A	MS B	MS C	IPR
	Start of technology development	Start of design	Construction authorized	Follow-on construction authorized
Late	MS A	IPR	MS B	MS C
	Start of technology development	Start of design	Construction authorized	Follow-on construction authorized
Variable MS B	MS A	MS B	MS C	IPR
	Start of technology development	Transition from Navy to contractor	Construction authorized	Follow-on construction authorized

held once testing of the first ship is complete and a decision regarding the construction of follow ships is required.

In these alternatives, the milestones do not necessarily signify a transition from one phase to the next, as currently defined by DoDI 5000.02. However, each alternative placement of milestones still relates to specific decision points in a program life cycle and can be usefully defined.

Moving the milestones earlier could make it more challenging to satisfy some statutory and regulatory documentation requirements. In the "early" construct, the technology that will be put on the ship and the design of the ship are still being developed. As a result, the data and information necessary for the required reports could be less technically mature. For example, an Independent Cost Estimate would be based primarily on engineering and design parameters if MS B occurs at the start of a design. However, the potential benefit of an earlier milestone is that the technology can be fielded more quickly, assuming that no major technical problems emerge.

If the milestones occur late, the risks associated with a compressed schedule are minimized. The technology and program baseline are more mature when decisions must be made, thereby reducing the production and other risks associated with committing early to a specific technology. Nevertheless, the amount of time it would take to complete construction of the first unit may not align with force structure requirements or address industrial base concerns, as there would likely be a substantial delay until testing of the first unit is complete.

The "Variable Milestone B" construct allows MS B to occur when the design of the ship moves from the Navy to a lead contractor. This can occur as early as the conceptual design or as late as detail design and construction.

Figure 5.6 uses these different milestone definitions to show three potential ship design/build and oversight processes. The first alternative shows a design/build process with no overlap and milestones occurring "late." The second alternative depicts a design/build process with overlap between design and construction and a variable

Figure 5.6
Three Ship Design/Build and OSD Oversight Process Alternatives (Unconstrained)

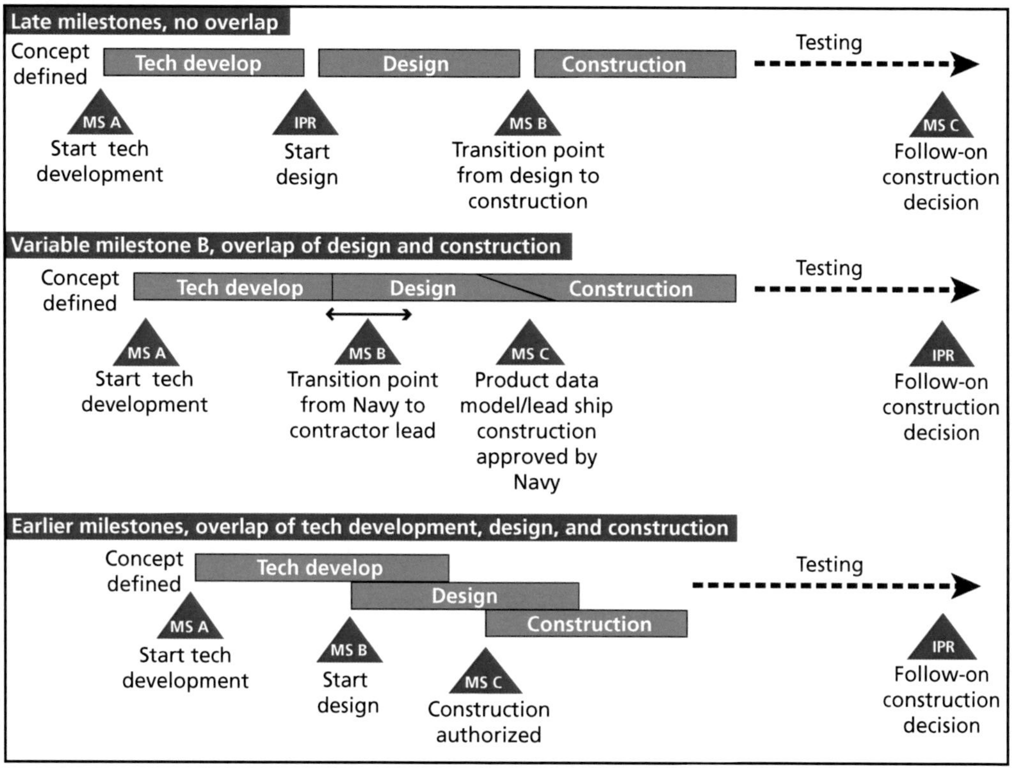

RAND *MG991-5.6*

MS B. The final alternative depicts a design/build process with overlap between tech-nology development, design, and construction, with earlier milestones.

In all of these unconstrained alternatives, no additional units are procured beyond the first until testing is complete. Although this minimizes technical and operational risk, it could pose a significant challenge to the industrial base.

None of these alternatives is the "right" alternative; each has pros and cons. In the "late milestones, no overlap" alternative, the technology and program baseline are more mature when decisions are made. Rework and redesign risks are minimized. However, longer program durations may also lead to requirements "creep" and pose a significant challenge to the industrial base. In the "variable Milestone B, overlap of design and construction" alternative, the milestones can be aligned with the key functional activi-ties of the ship design/build process. This option will require explicit process tailoring to define when MS B occurs for each ship program. The "earlier milestones, overlap of technology development, design, and construction" alternative would potentially allow mature technologies to be fielded more rapidly. These milestones mark the start of key functional activities in the ship design/build process. Nevertheless, satisfying the documentation requirements earlier in the process may be very challenging or might require a waiver. The concurrency of the process introduces risk of various kinds. The "optimal" program structure is thus very closely tied to the characteristics of a particular ship concept (e.g., technological maturity, design maturity, and relevant industry capabilities), and an acceptable balancing of associated risks.

In an unconstrained world, there are a number of desirable "best" practices. The shipyards, weapon-system contractors, and NAVSEA should collaborate, beginning with feasibility studies. Lead responsibility would shift depending on activity, life-cycle phase, and relative competency. Prototyping (with or without competition) would be done to the maximum extent possible at the component and subsystem level. Early and continuous developmental and operational testing would be performed. Verification through inspection, analysis, modeling and simulation, similarity, and demonstration would all be acceptable practices. The gap between lead ship and follow ship would increase to reduce technical and operational risk; the follow ship would be built only after IOT&E is complete.

However, the Navy must acquire ships in a world with a number of constraints. For instance, in the constrained real world environment, waiting to procure the follow ship until after the lead ship has completed IOT&E is considered impractical, resulting in an excessive production gap, learning loss, higher material costs, and vendor base impacts. The programmatic choices the Navy can make are influenced by these external constraints, which we refer to as variables.

Constrained Ship Design/Build and Oversight Options

The list below shows the key variables that constrain the ship acquisition process. These constraints influence the programmatic decisions the Navy takes:

- technical/engineering
- regulatory and statutory requirements
- industrial base parameters
- workforce
- shipyard financial viability
- capital equipment/tooling
- force structure requirements
- political factors
- fiscal constraints.

The increasingly seamless nature of the technical and engineering activities of the ship design/build process makes the placement of oversight activities, typically tied to the beginning or end of a development phase, particularly challenging. The standard development phases in the DoDI 5000.02 model do not necessarily correspond to the natural phases or sequence of activities in the ship design/build process.[4]

Regulations and statute dictate when certain activities must occur and can define what constitutes these activities. Table 2.1 and Table 3 in Enclosure 4 of DoDI 5000.02 summarize the many regulatory and statutory requirements. Fiscal constraints may affect the timing and scope of oversight activities. Alterations to funds allocation can delay development decisions. A number of other constraints also affect the timing and scope of oversight activities, including political influences, technical maturity of the system(s), force structure requirements, and industrial base considerations. Political influences can affect who, where, and when a program is produced. Technical maturity of the systems can determine what phase of development the program will enter and will be used to establish the level of technical review required. Force structure requirements can affect when certain activities occur, to maintain a certain level of capability. Industrial base considerations may also affect the timing of construction and contract award activities, if efforts to preserve industrial capability are required. All of these influences may affect the DoDI 5000 phase length and overlap or the extent to which there is a gap between the lead and follow ships.

When we apply these constraints to the various ship design/build and oversight processes, we can derive a common "constrained" process, as shown in Figure 5.7.

Depending on the design and manufacturing tools chosen, a complete separation of the design and build steps may be infeasible. Given the transition to design and manufacturing tools that promote a more seamless and integrated ship design/build process, that process is more likely to be highly concurrent than sequential, as depicted in Figure 5.7. A complete separation of development efforts and construction also does not seem feasible, given the conflicting program objectives that tend to maximize

Figure 5.7
Constrained Ship Design/Build and Oversight Process

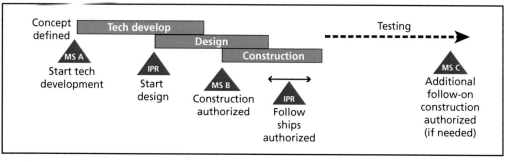

---

[4]   See Fireman, 2007, chart 4.

technology and minimize production time. Finding the right balance is a challenge for ship programs. A significant gap between the lead ship authorization and authorization of follow ships, which occurs on the completion of testing activities, would not likely occur in a constrained environment. Industrial base pressures would likely result in the need to contract for follow ships before the completion of IOT&E of the first ship. As a result, an IPR is likely to occur some time after MS B to authorize some number of follow ships, and MS C becomes a decision to authorize additional units.

We place Milestones A, B, and C to satisfy most existing regulatory and statutory constraints. DoDI 5000.02 specifies that Milestone B occurs before the first increment is produced and authorizes program initiation. For ships, this is typically before the start of construction of the first vessel. Milestone C typically occurs near the completion of IOT&E of the first unit, as depicted in Figure 5.7.

Summary

The continuous and overlapping nature of the design/build process for modern ships makes the placing and timing of acquisition milestones somewhat problematic. Technical information is continually evolving and may progress into early construction. As practiced today, these oversight points typically occur at major contract events, which may be the most logical place to pause and evaluate progress. However, this would lead to a sequential process that might result in workload gaps for the shipyards.

In our unconstrained oversight case, we examined a number of different timings for the acquisition milestones. Whereas MS A is relatively fixed as the start of major activity, MS B and MS C have some latitude. MS B may be as early as the beginning of design work or as late as the start of construction for the lead ship. Similarly, MS C could be as early as the start of construction for the first ship or as late as the completion of IOT&E. The choice of timing depends highly on the characteristics of the system. The main observation is that it is possible to define oversight points (milestones and technical reviews) that make sense for a given ship program and represent the functional equivalents and satisfy the intent of DoDI 5000.02.

The nominal process used for current ship programs does not align with any of the alternatives presented, although is closest to the constrained case.[5] However, there is one significant difference in that MS B denotes the start of detail design and authorizes lead ship construction, and an IPR authorizes initial follow ships. The best place for MS C is still unclear, particularly where the build quantity is small. SECNAVINST 5000.2D places MS C as we have done in the constrained case, as a continuing production decision. It is possible to move MS C earlier to replace the IPR that authorizes

---

[5] The major difference being that the start of design for the current nominal process begins with MS B and not an IPR. The construction authorization point aligns with an IPR.

the follow ship construction. This may better reflect the DoDI 5000.02 process, in which MS C begins initial production. Nevertheless, in our discussions with Navy and OSD stakeholders, we found mixed opinions regarding the movement of MS C to an earlier point in the process. Some felt that moving MS C earlier would provide only limited oversight value, because much information would have changed only marginally since the MS B review. Others thought that replacing the production IPR with MS C would be better, because it would formalize the real start of construction. We do not suggest a preferred or optimal placement of MS C for ships; we note only that its position should be a topic of early process tailoring.

CHAPTER SIX

Conclusions and Recommendation

Ship programs are subject to the same broad trends affecting other MDAPs, including rapid technology change, increasing system complexity, unit and program cost increases and associated affordability issues, fewer new program starts, and industrial base concerns. Nevertheless, ship programs also differ from other programs in many ways, including the size and complexity of the system, the length of time it takes to design and build a single unit, and the high unit cost. These and other characteristics of ship programs present unique acquisition challenges, including

- the timing and scope of Milestone B
- the timing and scope of Milestone C
- the application of LRIP/FRP decisions in relation to testing and phase exit criteria
- the reasonableness of completing IOT&E before production start
- the timing and scope of technical reviews, including PDR, CDR, and PRR.

These acquisition challenges concern questions of content at decision points and program reviews and the relative maturity of system design, technology, and production processes at those points in the program life cycle. Additional challenges deriving from these primary challenges include (but are not limited to) the application of competitive prototyping and MS A and MS B certification requirements.

In this chapter, we provide additional detail on the two aspects of the acquisition challenges for ships that we believe are the most important to address in any solution. We then outline the range of possible alternatives and provide more detail on one option that we believe is the most feasible and practical solution in the near term.

Two Aspects of the Acquisition Challenges Have Policy Implications

DoD acquisition policy has historically recognized the need to tailor some aspects of the acquisition process to meet the needs of a specific program. Ship programs are not unique in their need for process tailoring. Examples of tailoring include determining the program's entry point into the acquisition process, the timing of milestones and program reviews relative to contract awards, required information at milestones and reviews, and the timing and scope of test events. The last several revisions to DoDI 5000.02, including the May 2003 and December 2008 versions, have included language that allows tailoring. Accompanying guidance has suggested that such tailoring decisions be captured in the program's Acquisition Strategy, one of only four oversight documents requiring the signature of USD (AT&L).[1] Culturally and in practice, the acquisition community has recognized that the specific process tailoring needed may differ by commodity type or other program characteristics. And, as we showed above, program structures and characteristics may vary even within a commodity type, including ship programs.

The difficulties that stakeholders in the ship acquisition community (both OSD and Navy) have had in the past result from the interaction of several factors, including

- tension between tailoring the acquisition process and tailoring the program
- ambiguous language in policy and guidance documents
- differences in the organizational culture, norms, and interests (mission or responsibilities) of the different stakeholder offices.

Differences in organizational perspective resulting from agency mission and responsibilities, culture, and norms are generally not amenable to policy solutions. In fact, some such differences shape the governance structure, including differences between program management and oversight. Still, resolving issues of tailoring and ambiguous policy language are well within the realm of policy solutions and should be undertaken to improve the overall process. Unambiguous language, in both policy and guidance, to standardize tailoring would likely be helpful to stakeholders. Such language would set common terms and definitions (e.g., for milestones and program phases), set expectations, identify process or program characteristics that need to be considered as part of developing a program's Acquisition Strategy, and establish criteria for some tailoring decisions (i.e., when a program has certain characteristics, then a set of specific tailoring decisions applies).

Treatment of Tailoring in Policy and Practice

In the context of acquisition, the term flexibility is sometimes used as a synonym for tailoring. Defense Acquisition University (2006) describes flexibility as tailoring required program documentation, acquisition phases, and the timing, scope, and level of decision reviews. With the addition of certain testing issues, this definition captures the key dimensions of ship acquisition programs that need to be tailored in developing a suitable Acquisition Strategy.

---

[1] The four documents are the *Acquisition Strategy, Acquisition Program Baseline, Test and Evaluation Master Plan,* and *Acquisition Decision Memorandum.*

Tailoring tends to be easy to discuss but very hard to do. Part of the problem is confusion, or differences of opinion and interpretation, among stakeholders as to whether we are tailoring the *process* or the *program*. This confusion is exacerbated by seemingly conflicting language in DoDI 5000.02 (2008, p. 1). In the cover memo to the signed policy, paragraph 1c states that the Milestone Decision Authority has the authority to "tailor the regulatory information requirements and acquisition process procedures in this Instruction to achieve cost, schedule, and performance goals" (2008, p. 12). This explicitly allows the MDA to tailor the process to fit the needs of the program. In contrast, Enclosure 2, Procedures, paragraph 1b states that the "Program Manager (PM) and the MDA shall exercise discretion and prudent business judgment to structure a tailored, responsive, and innovative program" (2008, p. 12). This statement seems to suggest that it is the program that should be tailored.

In practice, of course, both the program and the process are tailored. This is often done ad hoc. The decisions and associated rationales are not fully recorded, and the stakeholders may not be clear whether a decision affects program structure or acquisition process. In general, we believe that process tailoring to fit the needs of the program will produce a more executable program. But it is important to note that tailoring the process does not necessarily mean that the basic principles of good acquisition policy and practice are lost. Rather, they are rearranged in a way that makes sense, given a specific program's characteristics. Some commodity types may require a relatively higher degree of process tailoring, including complex ships, satellites, launch vehicles, and software-intensive systems. Current policy explicitly recognizes the need for tailoring in ship programs, satellite programs, and programs containing large amounts of information technology.

A review of 17 approved Acquisition Strategies for 13 programs indicated that although most contain a paragraph on "tailoring and streamlining," few offered useful information. In particular, the full set of tailoring decisions and their rationale was not present in the approved Acquisition Strategy. This suggests that the DoD is missing an opportunity to make explicit the key process tailoring decisions and rationale associated with a specific program. Making those decisions explicit not only provides an important reference for stakeholders, but it also may help future stakeholders better understand program design and execution issues.

Explicit Treatment of Ships in Policy

Several acquisition policy documents recognize that ship programs may be different from nonship programs, and existing regulation does treat ships differently in terms of process requirements. We discussed these in some detail in Chapter Two (and in Appendix A), but here we review a few specific examples from the two most important policy documents: DoDI 5000.02 (December 2008) and SECNAVINST 5000.2D (October 2008).

DoDI 5000.02 allows for several process differences for ships, including

- Program initiation beginning at Milestone A. Program information reporting requirements would be adjusted accordingly.
- Lead ship authorization at Milestone B. Long-lead items for follow ships may also be authorized. Follow ship authorization is subject to MDA approval.
- LRIP defined as the "minimum quantity and rate that is feasible and that preserves the mobilization production base for that system."

These provisions appear to recognize differences in the timing of design maturity, concurrency of design and production (for the lead ship), and the relatively long construction duration for ships. It is interesting to note that DoDI 5000.02 does not say anything about ships with respect to Milestone C, FRP, or testing requirements—several of the most important ship acquisition challenges in need of policy resolution.

SECNAVINST 5000.2D also allows for process differences for ships, but it is not entirely consistent with DoDI 5000.02. It recognizes that

- Program initiation may be at MS A for ship programs (and reporting requirements are adjusted for these cases) and also explicitly allows that ship design activities may be concurrent with subsystem or component technology development.
- Milestone B authorizes lead and initial follow ships in quantity sufficient to sustain construction until an FRP decision.
- Milestone C and FRP may be combined into a single decision "for shipbuilding programs where follow ships are initially approved at Milestone B." The FRP decision "shall be held to provide the MDA the results of the completion of IOT&E," and authorize the construction of the remaining follow ships.

The concurrent ship design and technology development allowed in the Navy regulation is not mentioned in the DoD regulation but has historically been common practice. LRIP is not specifically tailored for ships. The most important difference is the expectation in the SECNAVINST that initial follow ships are authorized at MS B, whereas DoDI 5000.02 allows for long lead only for initial follow ships. This leads to a difference in treatment at MS C, which the SECNAVINST allows to be combined with FRP; the DoD regulation does not address FRP for ships at all. The Navy regulation appears to recognize that low-rate production begins with lead ship production, whereas the DoD regulation specifies that LRIP is part of the MS C decision. This results in greater ambiguity in the process requirements.

Policy Options

There is a range of policy options to address these two aspects of the policy problem: (1) tailoring program versus process and (2) ambiguous and conflicting regulatory language. From the least prescriptive (or most flexible) to the most prescriptive (or least flexible), these options include

- Exempt ship programs from DoDI 5000.02 entirely.
- Remove all references to commodity types in DoDI 5000.02.
- Permit ad hoc tailoring (current status quo).
- Clarify the language and interpretation for ship programs within DoDI 5000.02.
- Rewrite DoDI 5000.02 to include language for each commodity type.

Exempting ship programs from DoDI 5000.02 would give the Navy increased flexibility to design and manage ship programs. An extreme implementation of this option might exclude OSD oversight entirely. The Navy would still need to provide some degree of oversight; expenditure of public funds requires it. This option would therefore force the Navy to address the same set of issues: an ambiguous Navy regulation and balancing standardized oversight of weapon-system programs with the need to tailor processes to fit the characteristics of those programs. This option resolves the conflict between the DoD and Navy regulations, but the Navy would still need to ensure that its acquisition process regulation is unambiguous with respect to how ships are treated. This option parallels how national security space systems are treated—a commodity-based regulation at the Service/agency level.

DoDI 5000.02 recognizes the need to tailor some attributes of the acquisition process to better match the unique characteristics of specific programs. Tailoring guidance is provided at a very high level and is therefore somewhat vague and ambiguous. Ship and satellite programs are the only weapon-system types explicitly mentioned in the regulation. One option to address this problem is to remove explicit mention of commodity types in the regulation, leaving just the high-level guidance to tailor processes appropriately. At a practical level, however, this option does not address the core issues we have identified. In particular, stakeholders would still debate what program characteristics require process tailoring.

Ad hoc tailoring is essentially what takes place under current policy and practice. This allows process tailoring but it does not resolve any of the issues that make such tailoring problematic and challenging for stakeholders.

Clarifying the language and interpretation for ship programs in existing policy requires three separate but related activities:

- Clarify language in DoDI 5000.02 to resolve existing internal ambiguities.
- Align SECNAVINST 5000.2D with the revised DoDI 5000.02 to ensure that there are no conflicts. The Navy regulation can expand on the DoD regulation but should not contain any requirement that conflicts with it.
- Write a "policy memo" containing more specific guidance on what and how to tailor ship programs.

This option takes existing regulation and attempts to resolve ambiguities and conflicts in requirements. The additional guidance on ship program tailoring can ensure a more standardized interpretation of that regulation and offer specifics on program tailoring that reconcile the interests of program management and oversight officials. The policy memo makes explicit those aspects of the acquisition process that need to be tailored for ship programs and the range of tailoring options available given program characteristics.

Rewriting the foundational acquisition regulation to include language for each weapon-system program type is a highly prescriptive option for ensuring that the unique characteristics of different commodity types are explicitly accounted for in formal regulations. As such, this option reduces program management flexibility significantly. This option is equivalent to writing acquisition process regulations specific to each weapon-system type. The result would be different processes for ships; satellites; launch vehicles; missiles; armored vehicles; aircraft; and command, control, communications, computers, intelligence, surveillance, and reconnaissance programs. The main problem with this approach is that it will ultimately result in a number of completely different, independent acquisition processes, each with unique language, phases, and oversight requirements. An additional practical complication is determining the appropriate level of detail. Aircraft can be subdivided into tactical (small) and strategic (large) aircraft programs that can be very different from each other and therefore require different processes (and corresponding regulation). Similarly, ships can be subdivided into carriers, surface combatants, submarines, amphibious ships, and auxiliary ship programs, each with characteristics different enough to justify differences in process (and treatment in regulation). The end result would be a hugely complex set of acquisition regulations and corresponding processes that would significantly challenge both the program management and oversight communities.

Recommendation

We believe that clarifying the language and interpretation of existing regulations and guidance for ship programs is the most feasible near-term option and also strikes the right balance between prescription and flexibility for both program management and oversight officials. This solution has three parts, discussed below.

First, the language in DoDI 5000.02 needs to be made internally consistent and somewhat broader in scope to mitigate the most critical ambiguities. In the current language, MS B approves detail design and lead ship construction contract award, LRIP quantity, and advanced procurement for the follow ship. To better address the common ship program issues identified above, DoDI 5000.02 should also specify that LRIP may happen before MS C for ships, that IPRs approve exercise of the lead ship construction option in the contract (if needed), that IPRs approve initial follow ships under LRIP, and that MS C and FRP may be combined for ship programs. This last provision redefines MS C as FRP (rather than LRIP as is the current case) for certain ship programs.

Second, the language and intent of DoD 5000.02 and SECNAVINST 5000.2D must be aligned. The Navy's regulation can expand on concepts in the DoD regulation, but it should use the same language whenever possible and not add language potentially in conflict with DoDI 5000.02. In particular, SECNAVINST 5000.2D should use the same language for MS B decision scope, LRIP and IPRs, and MS C and FRP.

Last, additional and more specific guidance should be provided to ensure a more standardized interpretation of policy and process for tailoring. This guidance could take many possible forms, including a stand-alone policy memo or directive or a new chapter in the *Defense Acquisition Guidebook*. If a stand-alone memo, it could be signed jointly by USD (AT&L) and the Assistant Secretary of the Navy (Research, Development, and Acquisition) or their representatives (e.g., PSA/NW and DASN (Ships)). The guidance would make explicit the elements of the acquisition process for ship programs that require tailoring decisions and, perhaps if appropriate, constrain those decisions within an acceptable range. At a minimum, the acquisition challenges listed previously would be addressed, including

- the timing and scope of Milestone B
- the timing and scope of milestone C
- the application of LRIP/FRP decisions in relation to testing and phase exit criteria
- the reasonableness of completing IOT&E before production
- the timing and scope of technical reviews including PDR, CDR, and PRR.

Additional elements could be added, with stakeholder agreement, including how to handle competitive prototyping for ship programs and the MS A and MS B certifications required by statute. The idea is not to overly constrain a Program Manager's ability to tailor the process to meet program needs but rather to make all necessary decisions explicit so that key stakeholders can work together to reach them. The guidance can make explicit the expectations of oversight officials and provide clear criteria so that program officials can judge whether those expectations have been met.

Notional guidance for complex ships might include such steps as these:

- MS B approves award of a detail design and lead ship construction contract, advanced procurement for the lead ship, LRIP quantity, and an IPR date for exercising the lead ship construction option.
- Subsequent IPRs authorize follow ship construction in pairs.
- Live fire test and evaulation is waived, except for critical subsystems testable on dedicated test ships or land facilities.

- Development test and operating test(ing) are performed to the maximum extent possible at the subsystem level, including both land-based and test-ship-based testing when feasible. Test organizations assess results and make recommendations concurrent with lead ship construction.
- IOT&E is performed on the lead ship.
- MS C marks the end of development and occurs after IOT&E is complete. It will include FRP if the planned annual rate is greater than one, or simply continued production at the nominal rate.

An actual program is likely to require many additional decisions, not listed here, on the system engineering process and the timing and scope of technical reviews, integration of critical subsystems not managed by the ship program office, and the appropriate use of competition. The policy memo should be as comprehensive as possible in identifying the elements of the acquisition process that should be considered for tailoring while also establishing boundaries within which tailoring decisions are acceptable.[2]

None of these recommendations will have the desired effect on the ship acquisition process unless additional enabling actions are taken. The proposed solution works only if key stakeholders (both oversight and program management officials) agree to early and continuous interactions. Key oversight organizations and their responsibilities include PSA/NW (OIPT lead), Operational Test and Evaluation (OT&E) (test), SSE (system engineering), and ARA (Acquisition Strategy). PEO Ships and, of course, the Program Manager must also participate. The tailoring decisions this group makes, and the rationale behind those decisions, must be captured in the Acquisition Strategy approved at MS B and updated at subsequent milestones as needed.[3] This preserves those decisions in an authoritative source that will act as a reference for program execution. Last, the agreed Acquisition Strategy should be followed without significant deviation lest the entire set of tailoring decisions needs to be revisited.

---

[2]   In conjunction with this recommendation, we also recommend that a more thorough analysis of differences between ships and other MDAPs be undertaken, focusing on the development phase. Our analysis here suggests that there may be important differences in system engineering, design, testing, and technology development that could drive tailoring decisions.

[3]   A recent GAO report (2009b) is consistent with this recommendation.

Summary of Other Acquisition Documents Relating to Shipbuilding Programs

This appendix continues the overview of acquisition regulation from Chapter Two, where ship programs are treaded differently.

CJCSI 3170.01F (May 1, 2007)

This instruction describes the joint capabilities integration and development system. The document's focus is on the presystem acquisition phase—mostly requirements and needs definitions. There is only one reference to shipbuilding in the document relating to document formats:

> Where appropriate and with validation authority approval, mandatory documentation formats provided in reference c [the 3107.01 series] may be tailored to implement the intent of this instruction for specific programs, such as IT systems, shipbuilding, and national security space systems.

CJCSI 3170.01C (May 1, 2007)

This manual describes guidelines and procedures for implementing the Joint Capabilities Integration and Development System (JCIDS) process. It focuses on plans and staffing to generate the required JCIDS documents required by CJCSI 3170.01F.

> Guidance on the conduct of JCIDS analyses, the development of key performance parameters, and the JCIDS staffing process are provided in this manual. It also contains procedures and instructions regarding the staffing and development of joint capabilities documents (JCDs), initial capabilities documents (ICDs), capability development documents (CDDs), capability production documents (CPDs), and joint doctrine, organization, training, materiel, leadership and education, personnel, and facilities (DOTMLPF) change recommendations (DCRs).

Again, this document makes almost no reference to shipbuilding other than to acknowledge the early program initiation for shipbuilding programs at MS A through the approval of the CDD. "The CDD will be validated and approved prior to program initiation for shipbuilding programs."

Department of the Navy (*Acquisition and Capabilities Guidebook,* 2008)

This document is a companion to SECNAVINST 5000.2D and provides "discretionary" guidance and details for many areas of the acquisition process. There are no mandatory changes in the acquisition process for shipbuilding programs, just some implementation specifics. The following areas are where ship/shipbuilding is called out:

> One can use as the system supportability Key Performance Parameters (KPP) for CDD/CPDs Mission Capable/Full Mission Capable (MC/FMC) rates that are focused on primary mission areas (Annex 2-D, p. 27).

> Certain items are exempt from T&E provisions due to testing by others: Cryptographic or Cryptology equipment, Naval Nuclear Reactors and associated Systems, Nuclear Weapons, Medical and Dental Systems, Spacecraft and Space-based systems (Chapter 5, Integrated Test and Evaluation, p. 12).

- A T&E strategy for ships needs to include criteria to evaluate how (configuration, engineering, and functional) design changes would alter the appropriate level and scope of T&E required for increments of the lead and follow ships. The strategy should establish T&E requirements for both the ship and ship systems changes. Developmental test and evaluation (DT&E) and operational test and evaluation (OT&E) before Milestone B typically fills the role of T&E for individual shipboard systems. The T&E of individual weapon systems should use land-based test sites. For prototype or lead ship acquisition programs, T&E should be conducted on the prototype or lead ship as well as on individual systems (Chapter 5, Integrated Test and Evaluation, p. 15).
- For ship programs that do not include new development activities, and hence do not require OT&E, test at the shipyard and in various trials (e.g., builder's, acceptance, contract) may fulfill the TEMP requirements (Chapter 5, Integrated Test and Evaluation, p. 16).
- Live fire test and evaluation (LFT&E) for ships is mainly done through test and inspection to confirm that specifications for survivability (e.g., shock, fire, air blast) are being met. "During the 1-year shakedown period following delivery of the lead ship of a class, or early follow ship as determined in accordance with

reference (q), a full ship shock trial should be conducted to identify any unknown weakness in the ability of the ship to withstand specified levels of shock from underwater explosions" (Chapter 5, Integrated Test and Evaluation, p. 64).

- Systems integration is particularly challenging on ships compared with other weapons systems because of the great number of systems onboard. As a reflection of this complexity, the performance specifications for ships should also include interface definitions and interoperability characteristics. One focus is the topsides integration that provides connectivity to the rest of the fleet and mission system effectiveness. The suggested approach is a "systems engineering" process that balances all the competing factors. Examples of these factors are operability, interoperability, and safety and survivability. Integrated topside design should also seek to lower total ownership cost (Chapter 7, Systems Engineering and Human Systems Integration, pp. 11–12).

- The Navy also implements a Ship Characteristics Improvement Panel that provides guidance to the Resources and Requirements Review Board (Chapter 9, p. 11).

Defense Acquisition Guidebook (2009 Update)

This online guidebook (Defense Acquisition University, 2009) is the companion "document" (it formally resides on a Defense Acquisition University website) that provides specifics and details for the process put forward in DoDI 5000.02. It serves as a "how-to" and clarification guide for the general acquisition community and also ties to capture the intent of the regulation. The chapter headings are

1. Department of Defense Decision Support Systems (describes planning, programming budgeting, and execution; JCIDS; and defense acquisition processes)
2. Acquisition Program Baselines, Technology Development Strategies, and Acquisition Strategies
3. Affordability and Life-Cycle Resource Estimates (e.g., AoA, life-cycle-cost estimating, economic evaluations)
4. Systems Engineering (various technical deliverables such as requirements development, review documents, risk assessments, the "ilities")[1]
5. Life Cycle Logistics (planning, designing and implementing supportability)
6. Human Systems Integration (e.g., personnel, training, human factors, safety)
7. Acquiring Information Technology and National Security Systems

---

[1] Shorthand for reliability, maintainability, supportability, etc.

8. Intelligence, Counterintelligence, and Security Support (e.g., technology transfer issues, dual-use, security)
9. Test and Evaluation
10. Decisions, Assessments, and Periodic Reporting (describes all the decision points/gates, review boards, and the various deliverables and reporting requirements for programs)
11. Program Management Activities (a catch-all chapter for various PM activities, such as EVM.

There are a few specific references to ships in this guidebook, mostly referencing (and providing links to DoDI 5000.02):

- links to the parts of DoDI 5000.02 Enclosure 4 that list statutory and regulatory information requirements for ships at major milestones:
  - 7.3.6.1. Review of Information Support Plan (ISP)–Specific Mandatory Policies. Ships are mentioned with respect to ISP requirements for programs initiated before MS B. It is interesting to note that National Security Space Systems are mentioned in the same paragraph.
  - Other Contract Management Reporting. "Due to the extended construction process for ships, CCDRs are also required for the total number of ships in each buy and for each individual ship within that buy at three intervals— initial report (total buy and individual ships), at the mid-point of first ship construction (individual ships only) or other relevant time frame as the CWIPT [chair, working integrated product team] determines, and after final delivery (total buy and individual ships)."
  - 7.2.2. Mandatory Policies. "The table indicates that the Net-Ready Key Performance Parameter in the Acquisition Program Baseline, required at Program Initiation for Ships, Milestone (MS) B, MS C, and the Full-Rate Production Decision Review (DR) (or Full Deployment DR), in part satisfies the requirement. The table also indicates that the Information Support Plan (ISP), in part, satisfies the requirement. An Initial ISP is required at Program Initiation for Ships and at MS B. A Revised ISP is due at the Critical Design Review (unless waived). And the ISP of Record is due at MS C." Again, satellites are mentioned in the same paragraph as ships.
  - 7.5.13. Information Assurance (IA) Definitions. This section mentions surface ships as one component of an antisubmarine warfare mission to illustrate what is meant by Family of Systems.
  - 4.4.7. Environment, Safety, and Occupational Health (ESOH). "A current PESHE document is required at Program Initiation for Ships, Milestone B, Milestone C, and the Full-Rate Production Decision Review. It is recommended that the PESHE be updated for the Critical Design Review."

- A link to DoDI 5000.02, Enclosure 2 Section 7. "LRIP for ships and satellites is production of items at the minimum quantity and rate that is feasible and that preserves the mobilization production base for that system."
- A link to DoDI Enclosure 2 Section 6. "For shipbuilding programs, the required program information shall be updated in support of the Milestone B decision, and the ICE shall be completed. The lead ship in a class shall normally be authorized at Milestone B. Technology readiness assessments shall consider the risk associated with critical subsystems prior to ship installation. Long lead for follow ships may be initially authorized at Milestone B, with final authorization and follow ship approval by the MDA, dependent on completion of critical subsystem demonstration and an updated assessment of technology maturity."

The 2006 version of the DAG included these specific references to ships:

- System operating and support costs are normally expressed as annual operating and support costs individual system (e.g., ship).
- Disposal responsibilities. "The Chief of Naval Operations N43 and NAVSEA/ Supervisor of Shipbuilding act as managers for ship disposal and recycling" (Defense Acquisition University, 2006).
- Command, control, communications, computers, and information (C4I) Systems plans can take longer for ships. "Based on past experience with C4ISPs, for a small program with few interfaces, it takes about 6 months to get an ISP ready for a Stage I review. For most programs, ISP preparation for Stage 1 review takes about a year. For very complex programs, like a major combatant ship, it can take between 18 to 24 months. The process is based on development or existence of an architecture" (Defense Acquisition University, 2006).
- Because the development cycle is much longer for ships, PMs should involve testing authorities earlier in the process. "Naval vessels, the major systems integral to ship construction, and military satellite programs typically have development and construction phases that extend over long periods of time and involve small procurement quantities. To facilitate evaluations and assessments of system performance (operational effectiveness and suitability), the program manager should ensure the independent OTA [other transaction authority] is involved in the monitoring of or participating in all relevant activity to make use of any/all relevant results to complete OAs [operational assessmnts]" (Defense Acquisition University, 2006).
- Shock trails may fulfill the LFT&E suitability requirements. "Similarly, LFT&E tests such as Full Ship Shock trials might provide OT&E evaluators with demonstrations of operability and suitability in a combat environment" (Defense Acquisition University, 2006).

Federal Acquisition Regulation (GSA, DoD, and NASA, March 2005)

This document is much too large to summarize (it consists of nearly 2,000 pages, which include Parts 1 to 53). The document serves as the main regulation for acquisition in all the executive agencies. The Foreword states:

> The Federal Acquisition Regulation (FAR) is the primary regulation for use by all Federal Executive agencies in their acquisition of supplies and services with appropriated funds. It became effective on April 1, 1984, and is issued within applicable laws under the joint authorities of the Administrator of General Services, the Secretary of Defense, and the Administrator for the National Aeronautics and Space Administration, under the broad policy guidelines of the Administrator, Office of Federal Procurement Policy, Office of Management and Budget.

The primary objective of the regulation is to help the government procure items and services in a timely way while achieving "best value" and maintaining public trust and integrity.

Ship construction is mentioned multiple times, as is ship repair. The instances are mainly related to contractual issues. Here are the specific instances for new construction:

1. Subpart 14.2—Solicitation of Bids, 14.202.1. Shipbuilding acquisitions are exempt from preparing invitations to bid using the uniform contract format. However, the programs must still include relevant sections. The exception seems to be in place because several of the items are treated differently for ships (e.g., packaging and marking are not relevant).
2. Shipbuilding ceremonies, such as keel-laying, launch, and commissioning, are considered allowable public relations costs (31.205-1 e.5).
3. Solicitations must specify customary contract financing; for shipbuilding, progress payments are based on percentage or stage of completion (32.113).
4. Shipbuilding programs are exempt from the section titled Progress Payment Based on Costs—Subpart 32.5, when payments are based on a percentage or stage of completion (32.500).
5. Performance-based payments may not be used for shipbuilding contracts where progress payments are based on a percentage or stage of completion (32.1001 e.2).
6. The Contract Administration Office (CAO) may issue change orders and negotiate and execute resulting supplemental agreements for shipbuilding programs only if specifically authorized by the contracting office (43.302 b.8).
7. In terms of contractor liability for loss, a ship is considered a "high-value item" (46.802 2). This definition means that the government relieves the contractor from any contractual liability for loss or damage.

8. For Value Engineering Change Proposals (VECPs), the sharing period is extended because of the "prolonged" production period. The period is the later of either 36–60 months (at the discretion of the contracting officer) or the last scheduled delivery date of an item affected by the VECP based on the schedule at the time the VECP is signed. Furthermore, agencies may prescribe future savings under contracts awarded during the savings period (48-104-1 d). In such cases, the contracting officer must modify the contracting language to state all awards during the contracting period (48.201 h).

Defense Federal Acquisition Regulation (April 4, 2008)

The DFAR (DoD, 2008a) is the DoD implementation and augmentation of the FAR. It contains (i) requirements of law, (ii) DoD-wide policies, (iii) delegations of FAR authorities, (i) deviations from FAR requirements, and (v) policies and procedures that have an effect beyond those internal to DoD (from language in 201.301).

Specific instances where ships have specific or tailored regulations are listed below:

1. Ship critical safety items have different qualification requirements (Subpart 209.2). Such an item is defined as ". . . any ship part assembly or support equipment containing a characteristic the failure, malfunction, or absence of which could cause – 1) a catastrophic or critical failure resulting in loss of or serious damage to the ship; or 2) an unacceptable risk of personal injury or loss of life" (DoD, 2008). For these items, the contracting authority must obtain approval from the head of the design control activity before entering into a contract. The design critical authority for ships is the systems command that is specifically responsible for ensuring the seaworthiness of a ship or its equipment (this must be NAVSEA).

    NAVSEA is also the authority for approving any nonconforming safety critical item (246.407 S-70), although it may delegate this authority for minor nonconforming items.

    The design authority must concur with the issuance of a certificate of conformance for safety critical items (246.504).

2. For specific forging items (propulsion shafts, periscope tubes, and ring forgings and bull gears), procurement should be from domestic sources to the maximum extent possible (225.7102-1).
3. Vessel propellers must be manufactured in the United States unless they are commercial items or have a waiver (252.225-7023).

4. The construction of a vessel (or major component of the hull or superstructure) intended for the U.S. armed forces may not be awarded to a foreign shipyard, in accordance with 10 USC 7309 and 7310 (225.7013). There is also a similar prohibition against the overhaul and repair of vessels homeported in the United States (but the regulation does not apply to voyage repairs).

5. In accordance with the FAR, progress payments based on stage or percentage completion is authorized for shipbuilding programs (232.102).

6. A fixed-priced contract for a lead ship of a class may not be awarded unless USD (AT&L) approves (235.006).

7. USD (AT&L) must approve the prenegotiating position and final negotiated agreement for, and any increase of more the $250 million or any reduction of $100 million or more for, a fixed-price-type contract of a lead ship. USD (AT&L) must also approve an increase in the price or the ceiling price for contract of a lead ship of more the $250 million for equivalent quantities (235.006 b.ii.B).

8. For ship repair, nothing in a master agreement may relieve the contractor from complying with The Safety and Health Regulations for Ship Repairing (29 CFR 1915) or any other applicable federal, state, or local laws, codes, etc.

9. Other acquisition flexibilities—Subpart 218.1. "The contracting officer, without soliciting offers, may issue a written job order for emergency work to a contractor that has previously executed a master agreement, when delay would endanger a vessel, its cargo, or stores or when military necessity requires immediate work on a vessel" (DoD, 2008).

10. The procurement of welded anchor and mooring chain is restricted to U.S. sources except under specific circumstances (225.7007).

11. Among the clauses to include with solicitations, there are requirements for the "Inspection of Manner of Doing Work." These requirements deal with working to Navy specifications, using American Bureau of Shipping–qualified or DON-qualified welders, protection from fire, protection from freezing temperatures, and so forth. Such requirements/clauses are detailed in subpart 252.217. It is not clear whether these requirements apply to just repair and overhaul or also to new construction.

Additional Ship and Nonship Program Data

Table B.1
ACAT I Programs

Event	CVN-68	SSN-688	DDG-51	LPD-17	SSN-774	T-AKE	DDG-1000	CVN-21	LHAR	LCS	AMRAAM	C-17	SBIRS (High)	SM-6	F-22
MNS approved	Before 1965 if exists	Mar 70	Late 1970s	18 Sep 90	10 Oct 91	1992	26 Sep 94	Mar 96	31 Jan 01	Before 2003 if exists	No data	Dec 79	No data	17 May 99	No data
MS 0 approved	N/A	Nov 68	Feb 80	1 Nov 90	28 Aug 92	7 Dec 95	Jun 95	Mar 96	N/A	N/A	N/A	N/A	N/A	N/A	N/A
Preliminary design	No data	No data	Feb 82	Nov 93	No data	No data	17 Aug 98	No data	No data	July 03	No data	Early 1980	May 95	No data	No data
Design contract award date (system and/or lead ship)	Aug 65	Nov 69 and Feb 70	May 83	No data	FY 95 and FY 96	26 Aug 99	Nov 99	Oct 00	May 05	May 04	No data	Aug 81	May 95	No data	Oct 86
ORD or CDD approved	No data	No data	No data	8 Apr 96	Jun 95	23 Aug 99	Nov 97	Feb 00	8 Feb 05	Feb 03	No data	No data	15 Aug 96	No data	No data
MS I or MS A approved	Aug 65	Feb 70	Jun 81	11 Jan 93	18 Aug 94	26 Aug 99	12 Jan 98	15 Jun 00	20 Jul 01	May 04	Nov 78	Jul 82	May 95	N/A	Oct 86
PDR is held	No data	No data	No data	No data	No data	No data	Multiple PDRs: 2 in FY04, 1 in FY03	No data	No data	No data	Multiple PDRs	Multiple PDRs	Multiple PDRs	FY05	FY 06
MS II or MS B approved	May 68	N/A	Dec 83	17 Jun 96	30 Jun 95	18 Oct 01	23 Nov 05	26 Apr 04	Jan 06	June 10	Sep 82	Feb 85	Oct 96	Jun 04	Jun 91
LRIP approval date	No data	No data	30 Oct 86	N/A	30Jun 95	N/A	13 Feb 08	26 Apr 04	N/A	N/A	4 Jun 87	18 Jan 89	N/A	12 Jul 04	Aug 01

Table B.1—Continued

Event	CVN-68	SSN-688	DDG-51	LPD-17	SSN-774	T-AKE	DDG-1000	CVN-21	LHAR	LCS	AMRAAM	C-17	SBIRS (High)	SM-6	F-22
CDR is held	No data	No data	No data	No data	No data	May 02 and Apr 03	14 Sep 05	No data	11 Oct 05	N/A	Multiple CDRs	Multiple CDRs	Multiple CDRs	FY05	Feb 95
Lead ship detail design and construction contract award	31 Mar 67	No data	Apr 85	17 Dec 96	No data	13 Oct 01	8 Aug 06 (BIW) and 31 Aug 06 (NG)	10 Sep 08	15 Jul 05	15 Dec 04 (LM) and 14 Oct 05 (BIW)	No data	No data	No data	3 Sep 04	Aug 91
First follow ship construction contract award	29 Jun 70	Jan 71	May 87	Dec 98	Sep 97	No data	Sep 08	Sep 11	No data	26 Jun 06	May 87	31 May 96	Jan 08	3 Sep 04	Sep 01
PRR is held	No data	No data	No data	Jul 00	No data	No data	No data	Nov 08	11 Sep 08	No data	No data	No data	No data	No data	No data
MS III or MS C approved	Jun 70	Jan 71	Oct 86	Sep 09	Apr 09	Oct 01	Sep 15	Sep 18	N/A	FY 14	May 91	Jan 89	N/A	Sep 08	Mar 01
Lead ship delivery	11 Apr 75	2 Nov 76	29 Apr 91	20 Jul 05	12 Oct 04	20 Jun 05	Apr 13	Dec 15	Aug 12	18 Sep 08	Oct 88	Jul 90	Aug 2004; Dec 2009	Sep 09	Jun 03
OPEVAL starts	No data	No data	Jan 92	FY 08	FY 08	Nov 06	Sep 14	Oct 15	Sep 12	No data	No data	No data	No data	Aug 10	N/A
OPEVAL ends	No data	No data	Feb 92	FY 09	FY 09	Feb 07	No data	Aug 18	Sep 14	No data	No data	No data	No data	Sep 10	N/A
IOT&E starts	No data	No data	Jan 92	Jan 06	Jan 08	No data	Mar 15	No data	No data	No data	Oct 83	Dec 94	No data	No data	Apr 04

Table B.1—Continued

Event	CVN-68	SSN-688	DDG-51	LPD-17	SSN-774	T-AKE	DDG-1000	CVN-21	LHAR	LCS	AMRAAM	C-17	SBIRS (High)	SM-6	F-22
IOT&E ends	No data	No data	Feb 92	Oct 08	Oct 08	No data	No data	No data	No data	No data	Jan 89	Jun 95	No data	No data	Dec 04
First follow ship delivery	12 Sep 77	15 Jun 85	19 Oct 92	22 Dec 06	20 Jun 06	27 Feb 07	N/A	Sep 19	N/A	Approx. 09	No data	Jan 95	N/A	No data	Jun 03
Lead ship IOC is attained	No data	Nov 76	Feb 93	30 Apr 08	5 Mar 07	May 07	Mar 15	Sep 16	Feb 14	Jun 09	Sep 91	Jan 95	N/A	Sep 10	Dec 05
FRP approval or MS IV	Jun 70	Jan 71	Oct 93	Sep 09	April 09	Oct 01	Sep 15	Sep 18	No data	FY 14	Apr 92	Feb 96	N/A	Nov 10	Nov 06

Figure B.1
Littoral Combat Ship Program Structure

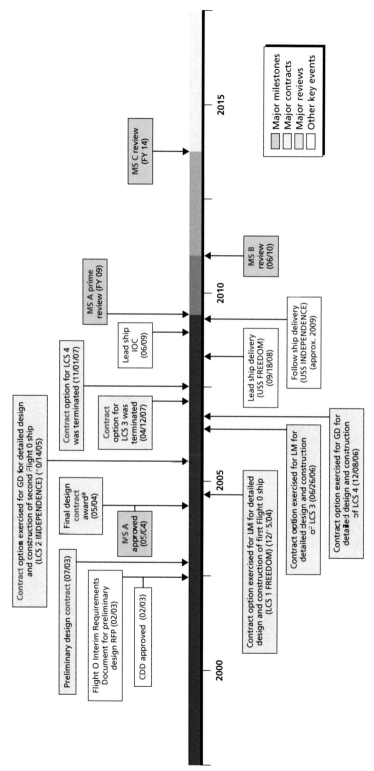

NOTES: Blue shading on time line indicates length of MS A, MS B, and MS C. No LRIP yet. No revised dates for OPEVAL or IOT&E available. No PDR and CDR information available yet. The LCS acquisition strategy was subsequently revised and dual block buys for 10 ships of each design (20 ships total) were competitively awarded in December 2010.

[a] On May 27, 2004, the Navy awarded contracts to two industry teams—one led by Lockheed Martin, the other by General Dynamics (GD)—to design two versions of the LCS, with options for each team to build up to two LCSs each.

RAND MG991-B.1

Figure B.2
LEWIS AND CLARK Class (T-AKE) Dry Cargo/Ammunition Ship Program Structure

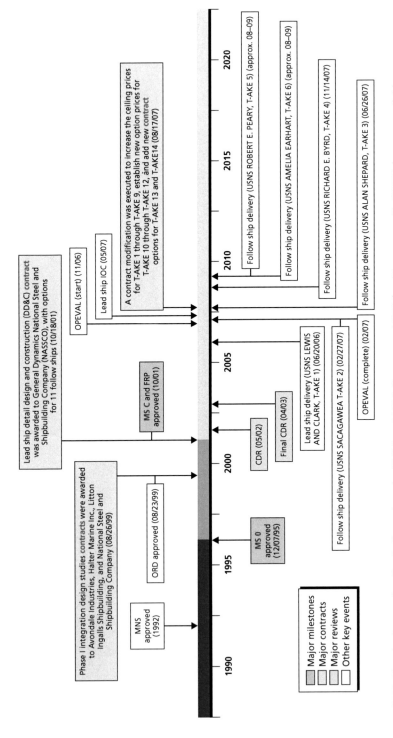

NOTES: Blue shading on time line indicates length of MS 0 and MS C. No approved LRIP for this program. No IOT&E dates are available.

RAND *MG991-B.2*

Figure B.3
Future Aircraft Carrier (CVN-21) Program Structure

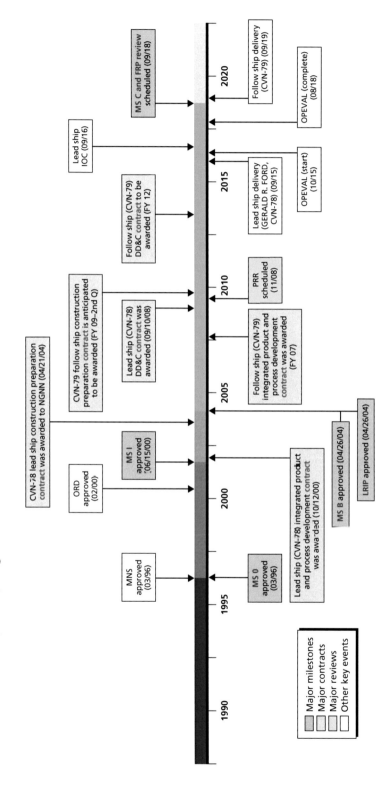

NOTES: Blue shading on time line indicates length of MS 0, MS I, MS B, and MS C. No CDRs were in the official or trade literature. No IOT&E dates are available.

RAND MG991-B.3

Figure B.4
Nuclear Aircraft Carriers (CVN-68 Class) Program Structure

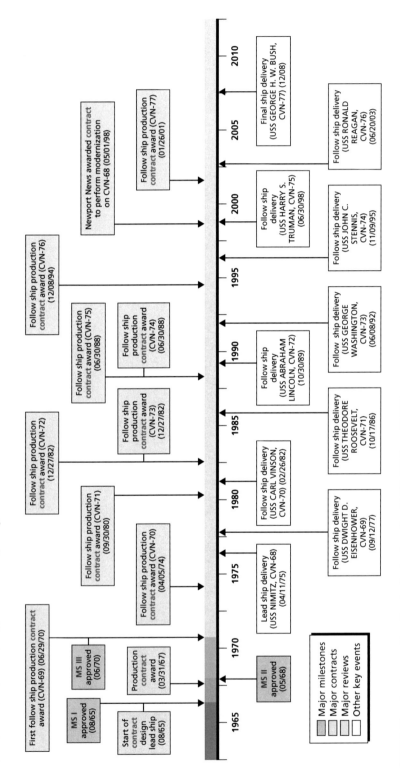

NOTES: Blue shading on time line indicates length of MS I, MS II, and MS III.

RAND MG991-B.4

Figure B.5
LPD-17 Class Amphibious Transport Dock Ship Program Structure

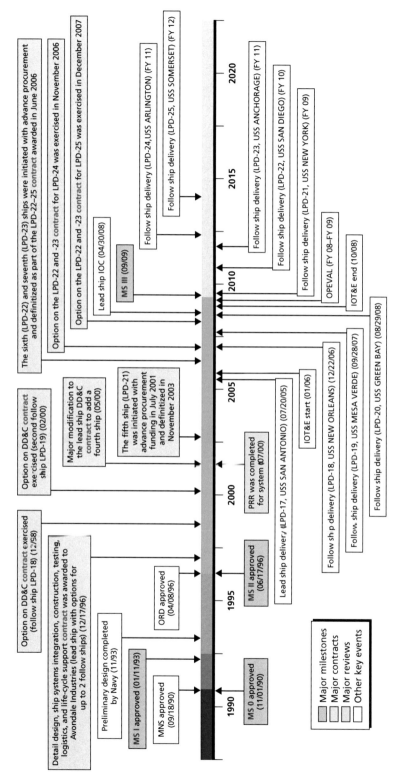

NOTES: Blue shading on time line indicates length of MS 0, MS I, MS II, and MS III. No FRP has been approved. No PDRs or CDRs were in the official or trade literature.

RAND *MG991-B.5*

Figure B.6
LHA Replacement Amphibious Assault Ship Program Structure

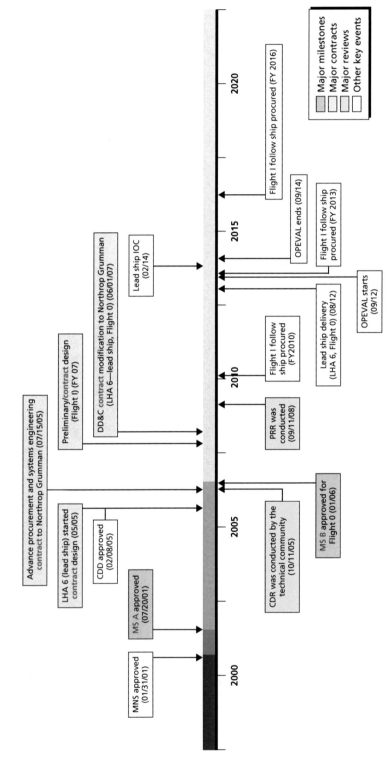

NOTES: Blue shading on time line indicates length of MS A, MS B. No LRIP has been approved. No MS C date has been approved or estimated. At award, the LHA 6 advance procurement contract was subsumed by the DD&C contract. No initial operational test and evaluation dates for Flight I were "planned" dates. No mention of actual dates in documentation.

RAND *MG991-B.6*

Bibliography

Assistant Secretary of the Navy (Research Development and Acquisition), Deputy Chief of Naval Operations (Integration of Capabilities and Resources), OPNAV 4000 Ser N8/7U62072, Washington, D.C.: Department of the Navy, July 19, 2007.

Birkler, John, Anthony G. Bower, Jeffrey A. Drezner, Gordon Lee, Mark Lorell, Giles Smith, Fred Timson, William P.G. Trimble, and Obaid Younossi, *Competition and Innovation in the U.S. Fixed-Wing Military Aircraft Industry,* Santa Monica, Calif.: RAND Corporation, MR-1656-OSD, 2003. As of June 18, 2010:
http://www.rand.org/pubs/monograph_reports/MR1656/

Castelli, Christopher J., "Audit Exposes Failed Management of Troubled Littoral Warship," *Inside the Pentagon,* January 31, 2008.

Cavas, Christopher P., "Secretary Wants Navy to Retake Shipbuilding," *Navy Times,* April 8, 2007 As of June 18, 2010:
http://www.navytimes.com/news/2007/04/defense_navy_shipbuilding_070404/

Chairman of the Joint Chiefs of Staff, *Joint Capabilities Integration and Development System,* Instruction, CJCSI 3170.01F, Washington, D.C., May 1, 2007a.

———, *Operation of the Joint Capabilities Integration and Development System,* Instruction CJCSI 3170.01C, Washington, D.C., May 1, 2007b.

Clark, Deborah L., Donna M. Howell, and Charles E. Wilson, *Improving Naval Shipbuilding Project Efficiency Through Rework Reduction,* Naval Postgraduate School Thesis, Monterey, Calif., September 2007.

Defense Acquisition University, *Defense Acquisition Guidebook,* November 2006.

Department of the Army, *Army Acquisition Policy,* Army Regulation 70-1, Washington, D.C., December 31, 2003.

———, *Army Acquisition Procedures,* Pamphlet 70-3, Washington, D.C., January 28, 2008.

Department of Defense, *Defense Federal Acquisition Regulation Supplement,* Revised April 23, 2008. As of June 18, 2010:
http://www.acq.osd.mil/dpap/dars/dfarspgi/current/index.html

Department of Defense, Under Secretary of Defense, Acquisition, Technology, and Logistics, *Operation of the Defense Acquisition System,* Department of Defense Instruction 5000.2, Washington, D.C., May 12, 2003.

———, Milestone Decision Authority Decision Forums, Joint Analysis Team, *Report,* July 2008a.

———, *Operation of the Defense Acquisition System,* Department of Defense Instruction 5000.02, Washington, D.C., December 2, 2008b.

————, Milestone Decision Authority Decision Forums Joint Analysis Team, *Approved Recommendations,* January 2009.

Department of Defense Inspector General, *Navy Acquisition Executive's Management Oversight and Procurement Authority for Acquisition Category I and II Programs,* Report No. D-2007-066, Washington, D.C., March 9, 2007.

Department of the Navy, *Acquisition and Capabilities Guidebook,* undated.

————, Assistant Secretary of the Navy, Research, Development and Acquisition, *Department of the Navy Acquisition Plan Guide,* Washington, D.C., March 2007.

————, Research, Development, and Acquisition, *Q&A,* March 2008. As of June 15, 2010: http://acquisition.navy.mil/acquisition_one_source/program_assistance_and_tools/ question_of_the_day/q_a_march_2008

————, Deputy Assistant Secretary of the Navy, Research, Development and Acquisition, *Department of the Navy (DON) Acquisition and Capabilities Guidebook,* SECNAVM-5000.2, Washington, D.C., December 2008b. As of June 18, 2010: https://akss.dau.mil/Documents/Policy/DON%20Acquisition%20and%20Capabilities% 20Guidebook.doc

Deputy Secretary of Defense, *The Defense Acquisition System,* Department of Defense Directive 5000.1, May 12, 2003.

Doerry, CAPT Norbert, *Systems Engineering, Knowledge, Skills and Abilities,* Naval Sea Systems Command, briefing presented at Engineering the Total Ship Symposium, September 23–25, 2008.

Drezner, Jeffrey A., Giles K. Smith, Lucille E. Horgan, Curt Rogers, and Rachel Schmidt, *Maintaining Future Military Aircraft Design Capability,* Santa Monica, Calif.: RAND Corporation, R-4199-AF, 1992. As of June 18, 2010: http://www.rand.org/pubs/reports/R4199/

Dur, Philip A., "Status of Shipbuilding Industrial Base," Testimony before the U.S. Senate Armed Services Committee, Sea Power Subcommittee, Washington, D.C., April 12, 2005.

Farr, John V., William R. Johnson, and Robert P. Birmingham, "A Multitiered Approach to Army Acquisition," *Defense Acquisition Review Journal,* April–July 2005.

Fireman, Howard, *Naval Ship Design Overview Briefing,* Washington, D.C.: Naval Sea Systems Command, July 19, 2007.

First Marine International, *Findings for the Global Shipbuilding Industrial Base Benchmarking Study, Part 1: Major Shipyards,* August 2005.

General Dynamics Electric Boat, *The Virginia Class Submarine Program: A Case Study,* Groton, Conn., February 2002.

————, *Design Build Overview: 21st Century,* Revision A, Groton, Conn., March 1, 2004.

General Services Administration, Department of Defense, and National Aeronautics and Space Administration, *Federal Acquisition Regulation (FAR),* Parts 1–51, Washington, D.C., March 2005.

Goddard, CAPT C. H., and CDR C. B. Marks, "DD(X) Navigates Uncharted Waters," *Proceedings,* U.S. Naval Institute, January 2005.

Government Accountability Office, *Actions to Improve Navy SPAWAR Low-Rate Initial Production Decisions,* Report to Congressional Committees, GAO-01-735, Washington, D.C., August 2001.

————, *Improved Management Practices Could Help Minimize Cost Growth in Navy Shipbuilding Programs,* Report to the Chairman, Subcommittee on Defense, Committee on Appropriations, House of Representatives, GAO-05-183, Washington, D.C., February 2005.

————, *Assessments of Selected Weapon Programs,* Report to Congressional Committees, GAO-08-467SP, Washington, D.C., March 2008.

————, *Defense Acquisitions: Perspectives on Potential Changes to Department of Defense Acquisition Management Framework,* Report to Congressional Committees, GAO-09-295R, Washington, D.C., February 2009a.

————, *Best Practices: High Levels of Knowledge at Key Points Differentiate Commercial Shipbuilding from Navy Shipbuilding,* GAO-09-322, Washington, D.C., May 2009b.

Heffron, John Sutherland, *The Impact of Group Technology-Based Shipbuilding Methods on Naval Ship Design and Acquisition Practices,* Cambridge, Mass.: Massachusetts Institute of Technology, master's thesis, May 1988.

Higby, John, *A Defense Acquisition University Interview with Admiral Charles Hamilton,* Noble Transcription Services, November 2007.

Muñoz, Carlo, "Young Issues New Guidance for DOD Acquisition Process," *Inside the Pentagon,* December 2, 2008.

————, "House Bill Signals Rising Support for DoD Acquisition Overhaul," *Inside the Pentagon,* March 19, 2009.

National Security Space Acquisition Policy: Interim Guidance for DoD Space System Acquisition Process, 2009.

Naval Air Systems Command, *NAVAIR Acquisition Guide 2006/2007,* September 20, 2006.

Naval Sea Systems Command, *Naval Propulsion Directorate, Report on Preservation of the U.S. Nuclear Submarine Capability,* Washington, D.C., March 3, 1992.

————, *Ship Design Manager Manual, Future Ship and Force Architecture Concepts,* Code 05D1, Washington, D.C., rev. October 30, 2006.

————, *Description of the Naval Ship Design Phases,* Briefing, Washington, D.C., February 2008.

Office of the Inspector General, *Acquisition of the AOE-6 Fast Combat Support Ship,* Audit Report, Report No. 92-030, Washington, D.C.: Department of Defense, December 27, 1991.

O'Rourke, Ronald, *Navy DD(X) and LCS Ship Acquisition Programs: Oversight Issues and Options for Congress,* Order Code RL32109, Washington, D.C.: Congressional Research Service, The Library of Congress, updated January 25, 2005.

————, *Statement Before the Senate Armed Services Committee, Subcommittee on Sea Power, Hearing on Navy Capabilities and Force Structure,* Washington, D.C.: Congressional Research Service, April 12, 2005.

————, *Navy Littoral Combat Ship (LCS): Background and Issues for Congress,* Order Code RS21305, Washington, D.C.: Congressional Research Service, The Library of Congress, updated August 18, 2006.

————, *Navy DDG-1000 Destroyer Program: Oversight Issues, and Options for Congress,* Order Code RL32109, Washington, D.C.: Congressional Research Service, The Library of Congress, updated April 11, 2008.

Peterson, Gordon I., *A Quest for Stability in Navy Shipbuilding Program,* interview with Michael C. Hammes, Deputy Assistant Secretary of the Navy (Ship Programs), Navy League of the United States, September 2000. As of June 18, 2010:
http://findarticles.com/p/articles/mi_qa3738/is_200009/ai_n8903418

Program Executive Office Ships, *DD(X) Program Office (PMS 500), Test and Evaluation Master Plan,* No. 1560, Revision C., DD(X) Destroyer Program, Washington, D.C., August 26, 2005.

Secretary of the Air Force, *Operations of Capabilities Based Acquisition System,* Air Force Instruction 63-101, Washington, D.C., July 29, 2005.

Secretary of the Navy, SECNAVINST 5400.15B, ASN(RD&A), Washington, D.C.: Department of the Navy, March 5, 2004a.

———, SECNAVINST 5000.2C, Washington, D.C.: Department of the Navy, November 19, 2004b.

———, SECNAVINST 5430.7N, Washington, D.C.: Department of the Navy, June 9, 2005.

———, SECNAVINST 5400.15C, ASN(RD&A), Washington, D.C.: Department of the Navy, September 13, 2007.

———, SECNAVNOTE 5000, DASN(RD&A)ALM, Washington, D.C.: Department of the Navy, February 26, 2008a.

———, SECNAVINST 5000.2D, Washington, D.C.: Department of the Navy, October 16, 2008b.

Sherman, Jason, "Young Battles Requirements Creep, Sizes Up Acquisition Challenges," *Inside the Pentagon,* February 5, 2009.

Simmons, L. D., *Assessment of Options for Enhancing Surface Ship Acquisition,* IDA Paper P-3172, Alexandria, Va.: Institute for Defense Analyses, March 1996.

Syring, CAPT James D., CAPT C. H. Goddard, CAPT M. Kavanaugh, and CAPT C. S. Hamilton, *System Engineering Plan (SEP) for the DD(X) Program, Milestone B,* March 31, 2005.

Syring, CAPT James D., and Susan Tomaiko, *DDG1000 Zumwalt Class Destroyers: Acquisition Strategy,* Program Executive Office, Ships, OUSD (AT&L), January 2008 (RESTRICTED).

Taubman, Philip, "Costly Lesson on How Not to Build a Navy Ship," *The New York Times,* April 25, 2008.

Tighe, Carla E., *An Industry Study of the U.S. Shipbuilding Base,* Alexandria Va.: Center for Naval Analyses, CRM 92-219, April 1993.

The Industrial College of the Armed Forces, *Final Report: Shipbuilding Industry,* Fort McNair, D.C.: National Defense University, Spring 2007.

Toner, Michael W., *Testimony Before the U.S. Senate Armed Services Committee, Sea Power Subcommittee,* Washington D.C., April 12, 2005.

Under Secretary of Defense, Acquisition, Technology, and Logistics, *Surface Combatant 21 (SC-21) Acquisition Decision Memorandum (ADM),* Washington, D.C., January 18, 1995.

Under Secretary of Defense, Acquisition, Technology, and Logistics, *Implementation of the Weapon System Acquisition Reform Act of 2009,* Directive-Type Memorandum (DTM)-09-027, December 4, 2009.

Winter, Donald C., *Remarks to Sea Air Space Exposition,* Washington, D.C., April 3, 2007.

Work, Robert O., "Small Combat Ships and the Future of the Navy," *Issues in Science and Technology,* Fall 2004.

Selected Acquisition Reports

ASDS, Selected Acquisition Report (SAR), Defense Acquisition Management Information Retrieval (DAMIR), RCG: DD-A&T(Q&A)823-381, December 31, 2005.

Cobra Judy Replacement, Selected Acquisition Report (SAR), Defense Acquisition Management Information Retrieval (DAMIR), RCG: DD-A&T(Q&A)823-365, December 31, 2007.

CVN-21, Selected Acquisition Report (SAR), Defense Acquisition Management Information Retrieval (DAMIR), RCG: DD-A&T(Q&A)823-223, December 31, 2007.

CVN-68, Selected Acquisition Report (SAR), Defense Acquisition Management Information Retrieval (DAMIR), RCG: DD-A&T(Q&A)823-161, December 31, 2007.

DDG-51, Selected Acquisition Report (SAR), Defense Acquisition Management Information Retrieval (DAMIR), RCG: DD-A&T(Q&A)823-180, December 31, 2007.

DDG-1000, Selected Acquisition Report (SAR), Defense Acquisition Management Information Retrieval (DAMIR), RCG: DD-A&T(Q&A)823-197, December 31, 2007.

F-22, Selected Acquisition Report (SAR), Defense Acquisition Management Information Retrieval (DAMIR), RCS: DD-A&T(Q&A)823-265, December 31, 2007.

LCS, Selected Acquisition Report (SAR), Defense Acquisition Management Information Retrieval (DAMIR), RCG: DD-A&T(Q&A)823-374, December 31, 2007.

LHA Replacement, Selected Acquisition Report (SAR), Defense Acquisition Management Information Retrieval (DAMIR), RCG: DD-A&T(Q&A)823-333, December 31, 2007.

LPD-17, Selected Acquisition Report (SAR), Defense Acquisition Management Information Retrieval (DAMIR), RCG: DD-A&T(Q&A)823-542, December 31, 2007.

SSGN, Selected Acquisition Report (SAR), Defense Acquisition Management Information Retrieval (DAMIR), RCG: DD-A&T(Q&A)823-337, December 31, 2007.

SSN-774 (Virginia Class), Selected Acquisition Report (SAR), Defense Acquisition Management Information Retrieval (DAMIR), RCG: DD-A&T(Q&A)823-516, December 31, 2007.

T-AKE, Selected Acquisition Report (SAR), Defense Acquisition Management Information Retrieval (DAMIR), RCG: DD-A&T(Q&A)823-592, December 31, 2007.